Engine Revolutions:

The Autobiography of

Max Bentele

SAE Historical Series

Published by:
Society of Automotive Engineers, Inc.
400 Commonwealth Drive
Warrendale, PA 15096-0001

Library of Congress
Cataloging-in-Publication Data

Bentele, Max, 1909-
    Engine revolutions: the autobiography of
Dr. Max Bentele/by Max Bentele
        p.   cm.
    Includes index.
    ISBN 1-56091-081-X
    1. Bentele, Max, 1909-   . 2. Mechanical
engineers--Germany--Biography. I. Title.
TJ140.B43A3 1991
621' .092--dc20   90-23361
[B]   CIP

To
Magda
and our children
Rose-Marie, Ursula and Brigitte

Foreword

Rarely does one experience the good fortune to read of major engineering achievements through the eyes of a person who not only participated in them, but whose contributions helped make them a success. Max Bentele's career has encompassed almost all forms of the internal combustion engine: reciprocating and rotary, two- and four-stroke cycle aircraft and automotive, and the gas turbine for both air and ground applications. Max is one of those unique individuals who tells his story from an insider's viewpoint with candor and humor. His witty insights of the many notable engineers and entrepreneurs with whom he worked or who crossed his path are given without antagonism.

The primary aim of <u>Engine Revolutions</u> is not to focus on the personal accomplishments of the author. Max's own efforts are told more to show how he and his colleagues worked to solve engineering problems, whether to discover why aircraft turbochargers failed in flight or how seals on the Wankel engine could be improved. The lessons he and his co-workers learned through their developmental trials and tribulations offer encouragement to engineers of later generations who suffer the inevitable failures encountered in perfecting any new design.

Possibly of even greater value is the opportunity for engineering students to better understand what their chosen profession is all about. Too often a technical paper or company publication has been "sanitized" to avoid any mention of the strenuous efforts and outright failures which so often occur during a product's birth and early years. In Max's story one hears the setbacks along with the successes. There is as well the interplay between people of differing ideas and philosophies. Although he uses terms and explanations that may sometimes be beyond the experience of the reader, the essence of what is described and the lessons to be learned are never left in doubt. His story's appeal is not limited to his peers!

Fortunately, Max could be persuaded to include more of his personal side along with professional involvements. Much of what makes this book compelling reading is how the world-shattering events through which he lived guided his career and affected his private life. However, what follows is, by choice, more a reflection on the accomplishments of striving engineers.

It was a pleasure for me to assist Max during the last stages of his writing and an honor when he asked me to add these few words. As an elected Fellow of the

Society of Automotive Engineers, Max Bentele typifies an SAE member who has enriched his Society's legacy. He and I are both indebted to SAE for publishing his story as a part of our Historical series. By means of this series, SAE preserves the heritage of its members and provides a way to relate how practicing engineers throughout the world created the technologies that we accept as essential to our daily existence.

Lyle Cummins
SAE Historical Committee

Preface

This book was started as a project of the SAE Historical Committee. I had given a verbal, *ad hoc* interview describing my professional career at the 1987 SAE Congress. A year later, the Committee Chairman, C. Lyle Cummins, Jr., discussed with me an expansion of this program. His plan was to document the history of the modern automotive technology in books to be prepared by SAE members and Fellows. I found that to be an excellent idea for a number of reasons.

Politicians, public figures, writers, artists and entertainers customarily write autobiographies upon their retirement; scientists and engineers seldom do so. Books on technology are sporadic; they deal mostly with organization and management rather than with engineering. The latter kind can be counted with the fingers of one hand or two. Examples are:

Sir Harry Ricardo (1885-1974) Memories and Machines: The Pattern of My Life, 1968; Theodore von Kármán (1881-1963) The Wind & Beyond, 1967, mainly aerodynamics; Robert Bosch (1861-1942), written posthumously by Theodor Heuss, 1946; The Memoirs of Ernest C. (Cliff) Simpson (1917-1985), Aero propulsion pioneer, sponsored by the Aero Propulsion Laboratory of the Wright-Patterson AFB, it was written posthumously in 1987; S.D. Heron (1891-1964) actively contributed to and wrote History of the Aircraft Piston Engine published by the Ethyl Corporation. Recently, other engine development histories were compiled and published; an excellent example is Internal Fire (19th Century IC engines) by C. Lyle Cummins, Jr., 1976 and 1988.

In contrast, most books on significant developments reflect second-hand information resulting in misconceptions and controversies. Daily, weekly and monthly publications report failures and catastrophes, while encyclopedias focus mostly on accomplishments.

With this lack of documentation, concern about problems is restricted to specialists. General awareness on a broad base, however, is necessary in order to address potential dangers before they lead to catastrophes. A case in point is the fatal crashes of the first commercial jetliner, the De Havilland Comet in 1953. Full-scale experiments traced the cause to fatigue failures in the fuselage skin. These catastrophes and expensive, comprehensive investigations were then reported worldwide. The Comet is hailed but its calamities are usually ignored in encyclopedias. No wonder that their reoccurrence after 30 years surprised all but those familiar with the history of aviation.

Lyle and I agreed that the SAE should pursue such vital history documentation. Then he surprised me with his invitation to write the first book of this series. I accepted, feeling highly privileged and honored.

I was blessed with the good fortune to have participated in the pioneering development of two new powerplants, the jet engine/gas turbine (1941) and the rotary Wankel engine (1958). I conducted my work in three different countries, under five governments, and in six companies, by fate not by design. To borrow Dean G. Acheson's words, "I was present at the creation," in the trenches as well as in the headquarters, experiencing both the agonies and ecstasies of the real world.

Since my college days, I was active in professional societies, disseminating my work and participating in debates on pertinent developments with my peers and the general public. I also tried to elevate the engineer's status in society.

Never considering my work as routine, I always searched for and applied the basic principles for each engineering task. Two problems drew my special attention, vibrations and gas leakages, each a *Leitmotif* throughout my career. They would hardly be known to laypeople but for recent headline-generating events: The metal fatigue cracks and failures of airliners and the leaking O-rings of a booster rocket which caused the tragic disaster of the space shuttle *Challenger* in 1986.

It is said that the most engaging powers of an author are to make *new* things *familiar*, and *familiar* things *new*. Although I was probably unable to achieve that quality, I did the best I could.

It is my wish to take this opportunity to thank all the fine people who have helped and advised me in my career. I am deeply grateful to them!

Apart from those mentioned in the book, I first recognize mentors and friends who have departed from this earth: Professor Dr.h.c. Wilhelm Buschmann (1886-1979), co-founder and publisher of *Motortechnische Zeitschrift* (MTZ); Professors Dr. Richard Grammel (1889–1964), Dr. Wunibald Kamm (1893-1966) and Dr. Karl Wellinger (1904-1976) of my alma mater in Stuttgart; the SAE Past Presidents Milton J. Kittler, Andrew A. Kucher, Leonard Raymond and C. G. A. Rosen, and the SAE Fellows Peter Altman, John Dolza, Sr., Paul H. Schweitzer and Arthur F. Underwood.

Then I express my sincere thanks to Lyle for his faith in me and his continual help and inspiration, and to the SAE Publications Group for their good work. Finally,

I think of all my other colleagues and friends who are still with us; they are too numerous to list. I hope they will enjoy reading this book and accept my sincere gratitude for their support and contribution.

To become an author instead of retiring puts a strain on family life. I am happy to report that my family was with me all the way. The written accounts of our lives in Germany, England and the United States revived many memories. My wife Magda, our daughters and their families provided valuable suggestions and assisted in putting the manuscript together. From the bottom of my heart, I thank them for their patience, encouragement and love.

Table of Contents

Introduction 1

Chapter 1. The Quest for Superior Specific Engine Power 3
            Single-Sleeve Valve Engines 3
            Flat-Disk Valve Engines 7
            Rotary Valve Engines 10

Chapter 2. The Last Hurrah of the High-Performance Aircraft Engine 13
            Exhaust Turbochargers 13
              Turbine Blade Failures 16
              Full-Admission Turbines 23
              Turbine Cooling 23
            High-Performance Aircraft Engines 26
              Supercharging and Turbocharging 26
              Otto-Diesel Engine 29

Chapter 3. The Dawn of the Jet Age 33
            From Turbochargers to Jet Engines 33
            Pioneer Jet Engines, He S 1 to He S 8 A 34
            Axial-Flow Jet Engines 41
              Wagner-Müller-Heinkel, He S 30 41
              Junkers Jumo 004 44
              BMW 003 47
            From the He S 8 to the Second-Generation He S 011 48
              He S 8 A 48
              He S 011 Development 50
                Compressor 52
                Combustor 57
                Turbine 57
                Jet Engine He S 011 61
            Propulsion Systems for Long-Range Aircraft 61
              ZTL DB 007 63
              PTL DB/He S 021 63
              ML He S 50 63
            Early Jet Engine Development - Postscript 64
            Appendix 3.1 — Turbine Rotor Blade Environment 73
            Appendix 3.2 — Jet Engine Firsts 75

Chapter 4. End of World War II 77
            Interregnum 77
            My Hiatus in the Western Occupation Sectors 80
            Appendix 4.1 — The Gas Turbine as a Vehicle Powerplant ... 85

Chapter 5. The Automotive Gas Turbine 87
            Journey to England 87
            The Tank Gas Turbine 89
            Thermal Shock Investigations 96
            Control System 99
            Industrial Gas Turbines and Other Events 99

Chapter 6. Stuttgart Again 105
            The New Heinkel Company 105
            OEM Engines 106
            Two- and Three-Wheeled Passenger Vehicles 108
            Gas Turbines, Aviation and Space 116
                Heinkel Jet Engine He S 053 122
                Curtiss-Wright Overture 123
            Appendix 6.1 127
            Appendix 6.2 — The Gas Turbine, Today and Tomorrow 129

Chapter 7. Aircraft Gas Turbines, Revisited 137
            Product Improvement Programs 137
            Market Life Extension 141
                Dieselization 141
                Muffling 141
                Propellertrain 142
            Turbine Cooling 146
                Convection Cooling 146
                Transpiration Cooling 147
            Study Engines 148

Chapter 8. The Rotary Engine Era 151
            The Wankel Engine 151
            Multifuel Rotary Engines 165
            Appendix 8.1 — Book Review 169

Chapter 9. Lightweight Aircraft Gas Turbine Engines 171
                Lift/Cruise Engines 171
                Supersonic Transport Engine 172
                Demonstration Engines 174
                Exotica 175
                    Liquid Metal Regenerator 175
                    Single-Rotor Lift Fan Engine 176

Toroidal Drive 177
Miscellaneous 177
Transition 179
Appendix 9.1 — MB Collection in Transportation History
Foundation of the University of Wyoming 183
Appendix 9.2 — Supersonic and Hypersonic Flight:
Research and Flight Milestones 185

Chapter 10. AVCO Lycoming 187
A Rotary Engine 187
Turboshaft Engines 189
Compressor Disk Failures 189
Tank Gas Turbine - Revisited 191
Automotive Gas Turbine Engines 194
Low-Power Gas Turbine Engines 195
Turbine Containment Failure 200
Retirement from Industry 201

Chapter 11. Post-Retirement Pursuits 205
Commercial Ventures 205
The Wankel Engine - Postscript 209
Rotary Engine Potpourri 211
Automotive Engines for Ecology and Fuel Economy 215
Steam Engines 218
The Stirling Engine 219
The Gas Turbine Engine 220
AGT 1500 in Abrams Tank 226
Garrett/ITI Gas Turbine GT601 228
Electric and Hybrid Vehicles 229
Energy Storage 231
Appendix 11.1 — The Wankel Fever as Expressed on the
Cover and in Headlines of Magazines 233
Appendix 11.2 — Comments on JPL's Study "Should
We Have a New Engine?" 237

Chapter 12. Engineering History, Collections, Rights 241
Appendix 12.1 — Presentation of the "Max Bentele
Collection" of Wankel, Rotary & Other Engines to the
Detroit Public Library's Automotive History Collection 245
Appendix 12.2 — The Max Bentele Collection 249

Biography 255

Index 263

Introduction

My Path to Combustion Engines

My birthplace, Jungingen, is situated a few miles north of Ulm on the Schwäbische Rauhe Alb (Swabian Rough Alb). The term "rough" refers to its blustery but healthy climate. Outsiders often apply this characteristic to the Swabians who live there, including me. Other traits of the Swabians are their frugality, sense of humor, and "Wanderlust" (roving spirit).

Born in 1909, I received my schooling in Jungingen and Ulm where I graduated from high school in 1927. Following a mechanics apprenticeship, I studied mechanical and electrical engineering at the Stuttgart Technische Hochschule (TH, Institute of Technology, now University) and graduated as Diploma Engineer in 1932. I was fortunate to obtain various jobs that provided the opportunity to further my education with doctoral work, first as an operation engineer at the Stuttgart radio station, then as a research engineer at the central radio laboratory in Berlin. During that time I met my future wife, Magda Pfister; in November 1937 we were engaged to be married.

My doctoral thesis, completed in 1937, specifically treated gas vibrations in pipelines. At that time, the field of noise abatement offered few job opportunities. Consequently, I tried to expand my work from the intake and exhaust system of internal combustion engines to the engine itself. In early 1938 this chance materialized with a position in the Instrumentation and Measurement Laboratory of the Brandenburgische Motorenwerke (Bramo), located in Berlin-Spandau, the town that ten years later became famous for its old prison. Bramo, a subsidiary of the Siemens Group, produced the then well-known small Siemens & Halske engines and had started to develop and manufacture larger engines.

That Bramo employment represented the real start of my career in the engine industry. (The Biography at the end of the book includes further details of my family background, education and life.)

There were two pleasant interruptions of my daily grind. My fiancée and I attended in February 1938 a show in the Deutschlandhalle, a new indoor sports arena. Hanna Reitsch (1912-1979), the famous aviatrix, entered the cockpit of a helicopter, Model FW-61, designed and built by Henrich Focke (1890-1979) and

1

powered by a Bramo SH 14a engine. Under her command the aircraft ascended vertically, hovered motionless in the air, moved ahead, sideways or in reverse, then flew circles over the audience and finally landed safely in the middle of the hall. This historic event, headlined in the press with "Flug im Saal" (Flight in a Hall), was an exciting adventure for both of us; it raised my enthusiasm for being active in the aircraft engine field. (A Bramo colleague, on loan to the Focke company, remarked to me, "a helicopter is not a flying but a vibration machine." At that time I did not imagine that 30 years later I would actively pursue the development of helicopter gas turbine engines and still have to solve vibration problems.)[1]

The other happening confirmed even more firmly that I made the right choice. I attended a Joint Meeting of the Deutsche Akademie and the Lilienthal Gesellschaft für Luftfahrtforschung (the German Academy and the Lilienthal Society for Aeronautical Research). More than 3000 participated, including 400 visitors from 24 nations.

My story is essentially presented in chronological sequence. In some instances, to avoid confusion, I chose to deviate from this rule to preserve continuity of the subject.

---

[1] To the best of my knowledge, flying a helicopter inside a building was never done again. In 1970 efforts were underway to fly one inside the new Houston Astrodome; however, the venture was finally abandoned.

Chapter 1

The Quest for Superior Specific Engine Power

Single-Sleeve Valve Engines

My fascination for unorthodox, novel and revolutionary engines began when Alfred H.R. (Sir Roy) Fedden (1886-1974), the venerable engineer from Great Britain, presented a paper[1] on the single-sleeve valve aero-engine which he had developed at the Bristol Aeroplane Company since 1926 and which was now ready for production.

Sleeve-valve engines were not new. The Daimler Motoren-Gesellschaft (DMG) in Stuttgart had produced engines since 1910 with the Knight double-sleeve for their automobiles which were famous for quiet operation, and to a lesser extent, for high oil consumption. At that time, they produced 10 percent more power than contemporary poppet-valve engines. In World War I, they also had been installed in military vehicles and had served well there, even with inferior lubricating oils. In 1919, peacetime production of cars was resumed with them, and also with a supercharged option.

Burt and McCollum had won motorcycle races with their single-sleeve engines in the mid-1920s. In order to achieve high specific output, Sir Roy based the Bristol engine on this design and adapted it for aircraft use. Two principal alternatives served this objective: (1) elimination of the exhaust valve whose high temperature induced detonation and restricted the engine compression ratio, and (2) replacement of the poppet valves and their operating mechanism with a single sleeve between piston and cylinder which opened and closed intake and exhaust ports located on the cylinder periphery.

---

[1] Sir Roy was the principal speaker at an international Joint Meeting of the Deutsche Akademie and the Lilienthal Gesellschaft für Luftfahrtforschung (the German Academy and the Lilienthal Society for Aeronautical Research) held in Berlin in May 1938.

I was particularly enthusiastic about the second item because my car's two-stroke engine operated in a similar manner. DKW (Das kleine Wunder), the car's maker, had coined for it the slogan, "only three moving parts" (piston, connecting rod and crankshaft). I appreciated its reliability and low maintenance. Furthermore, I had just tried to investigate and record the valve motion during opening and closing of a large air-cooled aero-engine for the purpose of optimizing the valve clearances and diminishing the high contact forces of the valve head and seat, the cam follower and the cam. The results of these measurements made obvious the sensitivity of the valve mechanism to the varying operating conditions. The single-sleeve valve sounded like a panacea for eliminating these problems entirely.

Sir Roy enlivened his talk with slides and a color movie. It was, I believe, the first color film I saw at an engineering session. One slide dramatically demonstrated the simplicity of the sleeve-valve engine. On its left side it showed the cylinder head, cylinder, piston and the sleeve which received with a ball joint, attached at its low end, an oscillating up-and-down and a rotary motion from a small crank. On the right side it showed the cylinder, piston and cylinder head with two poppet valves including their valve seats, double springs and retainers, rocker arms, pushrods, cam followers and the cam ring. The left side depicted a very simple arrangement indeed, in contrast to a complicated, cumbersome and heavy design on the right side.[2]

Apart from its simplicity, I saw great advantages in the mechanical design features of the sleeve-valve engine: ample port areas for intake and exhaust, their rapid opening and closing and low restriction of the gas flow, elimination of high stresses in the valve drive, and insensitivity to operating conditions and to service life. On the negative side, I had reservations about the heat flow from the piston and its rings through *two* oil films to the cooling medium on the outside of the cylinder wall, about the sleeve itself and about the severe space limitations of the cylinder head to accommodate two spark plugs *and* a fuel injector.

The distinguished audience overwhelmed Sir Roy with applause, but the German aircraft engine industry declined to enter into license agreements on his engine. The difficulty of direct fuel injection into the cylinder may have played a role. To the best of my knowledge, sleeve-valve spark-ignition engines were never offered with cylinder injection but only with single-point injection into the air-stream, now called TBI (throttle body injection.)

---

[2] This vivid display of simplicity vs. complexity was very impressive. No wonder that promoters of novel engines have repeated it with regularity. Examples are Rover and others with the automotive gas turbine, NSU/Wankel and Curtiss-Wright with the Wankel rotary engine, and recently a revived two-stroke engine from Australia. A typical picture of Bristol's sleeve-valve engine components is shown in Figure 1.1.

Fig. 1.1. Bristol sleeve-valve engine components (1938).

During World War II, my original enthusiasm for the sleeve-valve engine simplicity proved to be based on dubious premises. My inspection of a captured Bristol two-row radial engine revealed a bucket full of gear wheels for the sleeve drive. I believe there were over 100 gears!

In the mid-1950s, I had the privilege to learn more about the sleeve-valve engine history. It is filled with a combination of technology, secrecy, politics, alleged patent dominance, advertised successes, and claimed superiority.[3]

In the U.S., the Continental Motor Corporation, which was first to investigate the Burt-McCollum engine type, had a single-cylinder engine running in 1928 and subsequently dominated its development. An air-cooled seven-cylinder radial produced 400 horsepower. Other companies also experimented with single-cylinder engines. The measured performance was, however, hardly significant enough to warrant full development of this engine type. Progress with the poppet-valve engine and with aircraft fuels presented the sleeve-valve with the proverbial "fleeing target." Two items played a dominant role: (1) sodium-cooling of the hollow exhaust valve, and (2) increasing the fuel octane number with high lead content.

---

[3] During the development of the Wankel engine, I was working in an almost identical atmosphere. It is documented, with some true and some distorted facts, in a number of publications and books. The complete history of the sleeve-valve engine is still to be written.

Bristol took an extraordinarily long time for the final development of the engine. Success was at hand by 1939. Development of the poppet-valve engine had almost ceased by then and its factory was turned over to sleeve-valve engines. During the war years of 1939 to 1945, a vast number of air-cooled engines were produced, 14-cylinder two-row radial Hercules and 18-cylinder two-row radial Centaurus engines, the total amounting to over 57,000 engines. Liquid-cooled engines also were developed and manufactured, the 24-cylinder H-type Sabre by the Napier company and the Eagle by Rolls-Royce. These were the highest-power sleeve-valve engines ever in production.

After the war, Hercules and Centaurus engines were installed in both commercial and military aircraft. High standards of reliability, low service and maintenance requirements, long overhaul periods and low fuel consumption were achieved. However, fuel consumption was not low enough to permit nonstop transatlantic flights from the American east coast to central Europe. It was the Curtiss-Wright Turbo Compound TC 18 that achieved this milestone.

One of my friends, a patent attorney, advised, "An engine is fully developed when its production and/or its use is stopped." With this in mind, I compared the air-cooled sleeve and poppet-valve engines of the mid-1950s and arrived at the following conclusion. In the supercharged versions, the slightly higher fuel consumption of the poppet-valve engines was partially compensated by their lower oil consumption. Reliability, service and overhaul period were probably about equal; otherwise, the sleeve-valve engine would have captured a greater market share. Their power with air-cooling stayed below 3000 horsepower; turbocharging was not applied. I believe that they were not produced outside Great Britain and were mainly used by commercial and military operators who were in the sphere of British influence. On the whole, the Bristol, Napier and Rolls-Royce sleeve-valve aero-engines are a credit to the British design and manufacturing skills. Also, to the best of my knowledge, they present the sole non-poppet-valve four-stroke high-power engines that reached mass production in modern times.

In the late 1950s, the reciprocating engine in both its poppet-valve and its sleeve-valve versions succumbed to the gas turbine and jet engine for the propulsion of medium and large airliners. It had suffered the same fate earlier in the military sector.

The sleeve-valve engine history confirms the axiom, "To be successful in the engine (or any other) market, one has to be a bit better than the competition. To replace an existing engine entirely, one has to be a lot better." The sleeve-valve engine achieved the first objective but could not offer enough superiority to replace the poppet-valve engine in aviation and, I believe, in any other field.

Flat-Disk Valve Engines

Shortly after Sir Roy Fedden's presentation, I started work in the Applied Research Department of the Hirth Motoren Company in Stuttgart-Zuffenhausen, a Stuttgart borough. Hellmuth Hirth (1886-1939) had founded the company in 1931 for the development and manufacture of small air-cooled aero-engines for training and General Aviation aircraft. A unique feature was the employment of antifriction bearings throughout the engine. The crankshaft, patented and manufactured by his brother, Albert Hirth, was built up with Hirth couplings that rigidly connected the inner races of their undivided roller bearings with the crankshaft cheeks, thereby permitting material selection, heat treatment and manufacture best suited for each component. Lubrication to all engine bearings, the cylinder walls and the accessory drives was supplied by a small metering pump. This arrangement produces low friction losses and excellent cold-start capability.

All Hirth engines[4] used a common 1-liter inverted cylinder assembly in four- and six-cylinder in-line and V8 and V12 cylinder versions for the output range of 100 to 400 horsepower. They offered low cruise fuel consumption in the order of 0.5 lb/hp/hr and low weight, from 2.2 and 1.8 lb/hp of the naturally aspirated four- and six-cylinders to 1.7 and 1.5 lb/hp of the supercharged V8 and V12 engines, respectively. In the marketplace they were very successful and contributed to quite a few aviation prizes and long-distance record flights. One example, an Arado 79, powered by a Hirth four-cylinder engine, during a flight from Berlin to Sydney, Australia, covered the leg from Bengasi, Libya, to Gaya, India, 6400 km (4000 miles) nonstop, an international record at that time.

One of the objects of our department was to advance the engines in specific power, weight and frontal area. Following the trend of the times, replacement of the poppet valves was investigated. A design of a cylinder head, incorporating two counterrotating tubular sleeves for both the intake and the exhaust, produced the highest output and lowest frontal area, but on paper only. This project was interrupted.

On September 1, 1939, we were called to the cafeteria for an important announcement on the radio; Hitler had invaded Poland. The Second World War had started. The country, and with it Hirth-Motoren, was transformed to war status; air-raid precautions were instituted, as well as camouflages of buildings and interconnecting roads, and blackout and energy conservation measures. R&D engineers like myself were exempt from conscription; this status was reviewed at regular intervals. Initially, only a few employees were called for service in the Armed Forces. Privately, a more severe impact prevailed: no license for a private car; its tires confiscated; commuting by street car, train, bicycle or occasionally by bus; successive introductions of rationing of food and essentials.

---

[4] The 1988 Summer Olympics in Seoul revived a nostalgic memory. The emblem in the South Korean flag is identical to the Hirth-Motor trademark without the words.

Continuing with the valving investigation, an experimental single-cylinder engine suffered from valve scuffing and seizures which even ample lubrication could not prevent. Gas sealing was also found to be inadequate. These shortcomings, as a result of mechanical and thermal distortions of the sleeves, had been predicted by some engineers, including me. Our department head had to scrap this project in a hurry, apparently unauthorized for hardware procurement and tests. This and other misadventures finally led to his departure from the company. Subsequently, I was appointed Director of the Applied Research Department.

Afraid of another debacle, this time my own, I took a less original, more practical approach to the design of superior valve mechanisms. One such device, the flat-disk valve, already had reached an advanced experimental stage. Its origin can be traced to Felix Wankel (1902-1988), who became known worldwide in the early 1960s for his rotary engine. Since 1926, Wankel had conducted systematic studies and experiments on the sealing of the compressed air and combustion gases of reciprocating engines which led to sealing elements, "sealing networks" and, in the late 1930s, to the flat-disk valve engine. One particular sealing element which became known at that time was the Wankelwalze, a cylindrical retainer pin for sealing the gap of a piston ring. A similar device connects the ends of the side with the apex seals of the Wankel engine. I cannot recall having met with Wankel in person at that time.

In association with the Wankel Institute, Wolf-Dieter Bensinger (1906-1974), in the late 1930s at the Deutsche Versuchsanstalt für Luftfahrt (DVL),[5] Berlin-Adlershof, applied the flat-disk mechanism to an aero-engine, the Daimler-Benz 34-liter liquid-cooled inverted V12 engine DB 600.

The flat-disk valve arrangement is shown schematically in Figure 1.2. A flat disk located in the cylinder head contains two dual kidney-shaped apertures which, with the disk's rotation, connect with the intake and exhaust conduits of the engine. The combustion chamber is sealed on its valve side with a cylindrical sealing component whose flat face rubs against the disk, thereby obtaining a rotary motion to assist adequate lubrication. Its outer cylindrical face contains small piston rings to prevent leakage to the disk, which itself is sealed from the outside by concentric rings on each side.

Bensinger's flat-disk valve developments finally led to a prototype engine which was certified as Model DB 612. Production, however, remained in abeyance for a variety of reasons. Bensinger himself transferred in 1944 to Daimler-Benz (DB) in Stuttgart. After WW II, he became head of design for passenger-car engines and, in 1960, in view of his previous cooperation with Wankel, became head of DB's Wankel engine development.

---

[5] The DVL was equivalent to the Lewis Research Center (LeRC) of the National Advisory Committee for Aeronautics (NACA, predecessor of NASA), Cleveland, Ohio, and the Royal Aircraft Establishment (RAE), Farnborough, England.

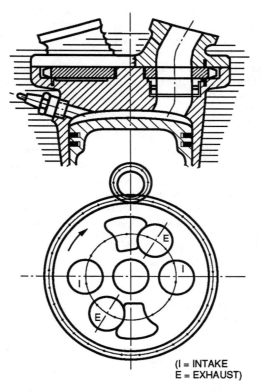

(I = INTAKE
E = EXHAUST)

Fig. 1.2. Flat-disk valve four-stroke engine (1939-40).

Our own analysis of flat-disk valves for four-stroke engines revealed some deficiencies: identical port areas for intake and exhaust, instead of favoring the intake; relatively slow opening and closing of the ports, thus diminishing the beneficial effects of the unobstructed gas passages; combustion chamber shape far from an ideal one. Figure 1.2 indicates the crevassed combustion chamber. The slow opening and closing relative to a piston-controlled port in the cylinder indicated poor performance. In view of these findings, we rejected this principle for the *four*-stroke and applied it to the *two*-stroke process where it could lead to an engine offering high specific output, low frontal area and acceptable fuel consumption.

Experiments conducted with a single-cylinder two-stroke engine, intake by piston-controlled ports in the cylinder and exhaust by a flat-disk valve in the cylinder head, confirmed the detrimental effect of the slow port opening and closing

9

on engine power which, in conjunction with the small port area itself, peaked at moderate engine speeds. To overcome this difficulty, we finally conceived a concept where *two* adjacent cylinders are operated by *one* disk. This arrangement also provided a compact combustion chamber. It was theoretically vastly superior to our present four-stroke engines in specific output, weight and probably also fuel consumption. Unfortunately, in view of other commitments, we were unable to verify our predictions experimentally.

Rotary Valve Engines

In addition to the flat-disk valve, we looked at other rotary valves. The best-known was the British design by the father-and-son team of Roland (1895-1970) and Michael Cross (1937-). A single cylindrical valve, located perpendicular to the engine cylinder axis, contains an intake and an exhaust passage separated by an integral wall. In the valve axis, the left outer end permanently connects with the intake, and the right end with the exhaust of the engine. During valve rotation, the peripheral openings of the intake and exhaust valve passages communicate in turn with an aperture in the cylinder head. The peripheral wall width thereby defines the degree of overlap between exhaust and intake.

The main advantages were: low valve temperatures, for which the term "Cross cool combustion chamber" was coined; ample port areas; absence of accelerations and decelerations in the valve train; and a compact combustion chamber shape. A 350 cc motorcycle engine converted to this valve mechanism achieved respectable performance with the high compression ratio of 10.5 and 66 octane gasoline. Various adaptations were pursued by Baer, BMW and others.

The also-British Aspin engine employed a conical valve, coaxial with the cylinder, part of which formed the whole combustion chamber with an opening to the valve cone. During valve rotation, intake, spark plug and exhaust alternately communicate with the combustion chamber. Again, low surface temperatures and no valve speed limitations were positive features of this design. It achieved gratifying engine performance.

Analyzing all of these rotary valve engines, we found on the negative sides gas sealing, lubrication and friction problems which would make them unsuitable for series production. Therefore, no further pursuit was warranted. Our experience with alternative valves showed that some designs may offer superior engine performance for specific applications but were uneconomical for the general market. We felt this design phenomenon may be best forgotten so that engineers could concentrate on more urgent tasks.

Was I wrong!

In the early 1970s, the Cross rotary valve was revived for the purpose of achieving high engine power with low-octane gasoline and also low exhaust emissions, both with passenger-car engines. An additional irony: The Esso oil company pursued this project at its research establishment in Abingdon, England, in association with Cross. To make a multi-cylinder engine feasible for series production, R.J.S. Baker conceived a cruciform arrangement of four cylinders, paired pistons, two common yokes, two eccentrics and a crankshaft for the power output. Baker designed the lower, Cross the upper portion of the engine which was variously called Cross-Baker, Abingdon-Cross or Esso-Cross. A 1.6-liter four-cylinder was expected to produce 140 hp at 6000 rpm for a family car, and a two-bank 3-liter eight-cylinder, 570 hp at 12,500 rpm for a race car. The four-cylinder was installed and road-tested in a Ford Cortina and Ford Consul.

It is said, "Old soldiers never die, they just fade away." It appears that nonconventional valve mechanisms will neither die nor fade away. In 1988, an American diesel manufacturer reported on investigations of a rotary valve for an advanced high-performance low-emission diesel for the next century.

Summarizing this century's valve mechanisms, the question arises why the poppet valve still maintains supremacy. In my opinion, it can be traced to two distinct positive features: (1) during the high-pressure phase, the poppet valve is *stationary* resulting in no gas leakage and no need for lubrication; and (2) its mechanism offers great flexibility in design and operation. The present vogue of multivalve engines and the advent of variable valve timing prove these points.

11

Chapter 2

The Last Hurrah of the High-Performance Aircraft Engine

Exhaust Turbochargers

As described in the previous chapter, the intake system, its valve mechanism, ports and timing can improve engine airflow and thus power, but to a moderate degree only. If much higher airflows and outputs are required, e.g., to lift greater aircraft weights at takeoff, to fly faster, to reach higher altitudes, or all of these, the ambient air has to be pressurized in a separate machine. Of the many types of compressors, aerodynamic superchargers are most suitable for aircraft. For the relatively low specific air consumption of piston engines, radial centrifugal types are preferred over axial-flow ones.

These compressors derive their driving power either from the engine itself or from a free turbine which is fed with the engine exhaust gas. The former are called gear-driven superchargers, the latter exhaust turbochargers or just turbochargers. Superchargers detrimentally affect the fuel efficiency somewhat; turbochargers raise it.

In the mid-1930s, layout and design of superchargers and their drivetrains were based on proven technologies, while turbochargers required new ones.

Exhaust gas turbines are subject to the full temperatures of the exhaust gases. The rotor blades of both axial and radial-inflow turbines present the severest design challenges: high centrifugal and gas bending stresses that occur at high metal temperatures. The exhaust gas temperatures of diesel engines are low enough for critical components to be made from heat-resisting steels or alloys. Those of Otto engines, however, were on the order of 1000°C (1832°F), too high for the then-available alloys. The blades and disk required some form of cooling.

The DVL explored the turbocharger field, both theoretically and experimentally, as did some aero-engine companies.

13

Hirth-Motoren was in the process of complementing its engine business with turbochargers. A prudent way for a small company to become active in this new engineering sector was to take a manufacturing license, thereby learning the trade and finally developing and marketing its own product. Our president, Curt Schif (1905-), did just that. It was the right time in view of the rising importance of military aircraft.

DVL's Institut für Strömungsmaschinen (Institute for Turbomachinery), headed by Professor Dr. Werner von der Nüll[1] (1909-) was furthest along with running hardware. He and his associates had performed work on the aerodynamic and mechanical aspects of all turbocharger components, radial-flow compressors in one- and two-stage versions, high-speed antifriction bearings, and turbines both with axial and radial-inward flow.

For turbine cooling, von der Nüll chose an axial turbine with solid blades and partial admission, one sector for exhaust gas and one for cooling air. It appears that a radial turbine was ruled out due to the formidable difficulties of cooling its turbine wheel. His prototype turbocharger presented a minimum-risk design with the following features: a two-stage radial compressor with shrouded impellers, guide vanes in the first diffuser with following passages to the second impeller, vaneless second diffuser with a spiral scroll for the air discharge; compressor rotor supported by a roller and a ball bearing; axial single-stage turbine with 50/50 admission of exhaust gas and ram air; separate exhaust gas housing with heat-insulated walls; turbine rotor supported by sleeve bearings in spherical outer housings; and a quillshaft connecting the compressor and turbine rotors. This construction eliminated adverse effects of misalignments and thermal distortions.

Hirth-Motoren obtained a few of these turbomachines for evaluation in a configuration which took full advantage of the turbocharger benefits. A 35-liter V12 aero engine, without its own gear-driven supercharger, fed its exhaust gases with two pipes fully or, by means of two wastegates, partially to the turbine. The compressor supplied pressurized air to the cylinders which were equipped for direct fuel injection. The setup in a test cell included full instrumentation and a test propeller to absorb the engine power. Figure 2.1 depicts the turbocharger installation.

Steady-state operation at the full ranges of engine and turbocharger speeds proved uneventful. All measured engine and turbocharger data agreed reasonably well with the predictions. After some slow accelerations and decelerations, we wanted to simulate a critical maneuver, a through-start following an aborted landing (wave-off). The engine running at "flight idle," the throttle was pushed fast to "wide open." We expected a sudden burst of power but looked at each other in disbelief! First, nothing had happened at all. Then, the engine and turbocharger picked up speed, ever so slowly. We had experienced the now famous "turbo lag!"

---

[1] Dr. von der Nüll and his institute were renowned for their success with analytical, design and experimental investigations of small radial compressors and radial inflow turbines.

Fig. 2.1. Turbocharger installation on a test stand (1942).

For modern automobile engines, turbocharged in this fashion, the turbo lag is still annoying and the focus of improvement efforts worldwide. Ceramic turbine rotors, variable turbine inlets and other means alleviate the problem but an economical solution has yet to be found. A satisfactory remedy for high-power aero-engines at the time was to employ turbocharging only in conjunction with an engine-driven supercharger.

Turbocharger development at Hirth-Motoren followed an evolutionary pattern. Advances in compressor technology permitted single-stage designs for pressure ratios, which, in conjunction with the engine supercharger, provided full manifold pressure up to altitudes of 12 km (40,000 feet). The very cautious design of two separate rotors as well as the bearing configuration were simplified. With these improvements, mass production was feasible and commenced at a slow but increasing rate. A serious problem, however, put a temporary stop to production.

Turbine Blade Failures[2]

In the blue summer sky of 1942, a twin-engine Heinkel He 111 aircraft cruised around Stuttgart at an altitude of 11 km (36,000 feet), higher than standard because its Junkers supercharged V-12 engines Jumo 211 were equipped with DVL/Hirth turbochargers to be flight-tested. A routine measurement indicated a slight drop in speed and air pressure of one of the turbochargers; however, its vibration and oil temperature levels were normal. Interruption of the flight was unwarranted but inspection after landing was in order. Lo and behold, two turbine blades had broken off at diametrically opposite locations. The small thermodynamic change and the regular mechanical behavior of the rotor were easy to explain, and the blade fractures probably the result of material defects.

One of the next flights took a less fortunate course. The speed of one turbocharger dropped suddenly to zero and made landing advisable. Again a blade loss, this time only one. The resulting rotor unbalance had caused seizure of the turbine bearing. This unfortunate sequence of events then occurred fairly regularly.

Inquiries at the DVL, the licensor of the turbochargers, provided little comfort and assistance. During their test and development programs they had *never* experienced any blade failures; ours were probably caused by our manufacturing process, material and its grain size, surface finish and general machining tolerances, they told us. In view of the ongoing mass-production of the turbochargers, our Hirth management became rather nervous, to put it mildly. Desperately searching for a way out of his dilemma, Schif took the unusual step of offering a financial award to anyone who would contribute to an answer to this formidable problem.

I myself vigorously started to investigate the blade failures, less for money than for prestige. Somewhat familiar with gas oscillations in pipelines, I suspected blade vibrations and set a target: to find their cause and to eliminate them.

The turbocharger incorporated a radial compressor and an axial turbine with solid rotor blades. The high exhaust gas temperature of the highly supercharged Otto-type engines necessitated turbine cooling. Von der Nüll had evolved a cooling scheme with partial admission of the exhaust gas and of the cooling air to the turbine. This design was at the time ahead of air-cooled hollow blades and, thus, first to be mass-produced.

The DVL/Hirth turbochargers were built in two models, 9-2216 with the ratio 50/50 for gas and air admission, and 9-2281 with a 60/40 ratio.

---

[2] Excerpts published in <u>VDI Nachrichten</u>, 25.11.1983.

The pressure of the exhaust gases before the turbine was much higher than that of the cooling air; the latter was ram air and its sector acted similar to a fan. Intuitively, I imagined that the sudden change from the strong gas force to the weak air force, and vice versa, would cause periodic excitation of the blades. If an exciting frequency would coincide with a natural blade frequency, i.e., under resonance conditions, the blade would vibrate like a tuning fork. The low aerodynamic and material damping would lead to high vibratory amplitudes and stresses and, in combination with the steady-state stresses caused by the gas and centrifugal forces, finally to a fatigue crack and blade loss. This rudimentary analysis clarified the causes of the blade failures, as well as the methods for their elimination. My task, therefore, was to treat the problem with mathematical and aerodynamic means and then to confirm it with experiments.

The turbine rotor blade attachment to the disk was a fairly rigid firtree. Hitting the blade airfoil with a soft instrument produced a distinct sound whose frequency was difficult to measure; a microphone and frequency analyzer were not immediately available. Musicians in my group came to the rescue. Stroking the airfoil tip with a violin bow generated the natural bending mode, stroking the trailing edge, the first torsional mode. To determine their frequencies, a piano was used; with a good ear and some experience, the corresponding key and thus frequency could easily be found. A violinist and a pianist were able to establish the frequencies of the 54 blades in a short time. (The fundamental bending frequencies of turbocharger blades were in the range of 2000 Hertz; those of jet engine turbines between 500 and 1000 Hz.)

I approximated the change of the blade frequencies under operating conditions: centrifugal forces cause an increase of the bending frequency due to the stiffening effect on the root attachment and on the airfoil; the higher blade material temperatures result in a decrease due to the lower Young's modulus of elasticity.

The gas and air flow sequence was simulated with two rectangles, with 0.95 amplitude for the gas force and minus 0.05 for the air. The Fourier analysis of the force sequence by means of formulae from mathematics books, though tedious and time-consuming, produced the individual harmonics, their orders and their amplitudes, the latter as a percentage of the gas forces.

For the 50/50 admission, only odd-numbered harmonics occur and their amplitudes diminish with their order number. Similar overall decline exists with 60/40 admission whereby all order numbers occur except the multiples of five. Their absence causes a slight amplitude increase of the neighboring order numbers.

To determine the resonance conditions, I constructed a frequency/speed diagram (Figure 2.2). It placed the speed range critical for blade vibrations as follows: For the turbocharger with 60/40 admission, the 6th and higher orders were in the operating range (as the diagram shows), for 50/50 the 5th order was above the maximum operating speed of 20,000 rpm, the 7th and higher orders were within.

As soon as these numbers became known, I overheard the following comment: "Bentele is one of our best engineers. Too bad, he now slipped into fantasyland! Seventh harmonic. What could that be?" But nobody, inside or outside the Hirth Company, had a better idea.

To corroborate my theory, I went a step further. I calculated the approximate danger level of the critical speeds from the turbine power, the gas and centrifugal force of a blade, and the exciting force from the Fourier results. Figure 2.3 depicts the conditions for 60/40. The 6th harmonic is more than twice as dangerous as the higher ones due to the prevailing higher gas, centrifugal and exciting forces.

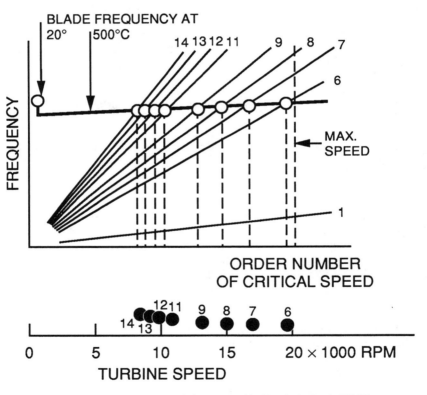

Fig. 2.2. Frequency/speed diagram (60/40 admission) (1942).

Fortunately, the 6th was near the maximum speed and could easily be placed above it by raising the natural blade frequency, initially by shortening the airfoil and finally by increasing its taper.

For the 50/50 model, the 7th harmonic was by far the most dangerous one. In addition, it occurred for a long time during climb of the aircraft at a turbocharger speed which could not be avoided. Therefore, on my suggestion, this model was discontinued immediately.

The results of my calculations and investigations put me in the admirable position of being able to explain the observed blade losses from the flight times spent in climb and cruise, and to build a scientific base for the means to eliminate these types of blade fatigue failure. I was satisfied with the findings and with myself, but still somewhat angry about the cavalier attitude of the DVL people who had tried to ignore the possibility of blade failures of *their* turbochargers and to put those experienced in flight operations entirely at our doorstep. I felt it needed a drastic demonstration to convince these and other doubters.

In a phone call to DVL, I explained my dilemma (with tongue in cheek) to von der Nüll: "Turbochargers manufactured in our machine shop suffered from blade failures; his original DVL turbochargers had never experienced a failure. Would he please lend me one of his turbines so that I could explore the differences and lay the problem to rest?" He agreed to my proposal somewhat reluctantly. Knowing me pretty well, he may have anticipated an unfavorable turn of events.

On receipt of the DVL turbine with 50/50 admission, we determined the natural bending frequencies of the 54 blades with the piano method and established its frequency/speed diagram. The blade frequency scatter put the dangerous 7th order in the speed range of 17,500 to 18,500 rpm, that of the less dangerous 9th in 13,300 to 14,100 rpm.

Our Test Department operated a "combustor stand" for turbocharger testing, allowing more freedom for the selection of individual parameters. The aircraft engine was thereby replaced with an electrically driven compressor and a combustor. To prove my theory, we established the following test program on the combustor stand for the DVL turbine. Run 1: One hour slowly sweeping the speed range 15,500 to 16,500 rpm, i.e., between the 9th and 7th harmonic, blade excitation without resonance. Run 2: One hour sweeping 19,500 to 20,500 rpm, above 7th, again free of vibratory resonances but with high centrifugal and gas forces. Run 3: Sweeping the critical speed range of 17,500 to 18,500 rpm.

Runs 1 and 2 proceeded uneventfully per plan. After one hour of Run 3, I received the call, "Nothing happened. Your theory is wrong!" I replied, "Theory's correct. Continue the run." After six more minutes, the call came, "You're right, a blade broke off!" *Quod erat demonstrandum.*

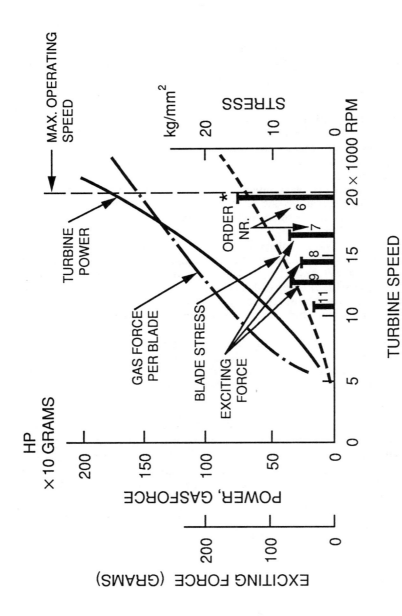

Fig. 2.3. *Danger level of excitation (60/40 admission) (1942).*

Figure 2.4 shows the turbine wheel and broken blade. Vibrations had caused a fatigue crack from the leading edge halfway across the airfoil. The remaining section was unable to carry the centrifugal load of the outer blade portion and broke in tension. The fatigue portion showed the typical seashell pattern, the rest that of a tensile fracture. Two more blades experienced fatigue cracks; their location at the leading *and* at the trailing edge puzzled us at first. We subsequently traced this anomaly to tolerances in the stacking of the airfoil sections.

This turbine wheel, with one broken and two cracked blades, provided the final convincing evidence for my theoretical and experimental investigations, and for my recommendations. The Hirth management gladly accepted them, put them into effect, and rewarded me with the lion's share of the prize money.

That wheel also helped to alleviate the strain in our relationship with our licensor, the DVL. It was done at a dramatic conference with the major participants, Professor von der Nüll, Schif, and me. Von der Nüll repeated the old arguments about the differences between DVL and Hirth in the manufacture of the turbine, grain size of the material, surface finish and so on. Schif asked, "Does that make your turbines immune to blade failures, Professor?" After the affirmative answer, Schif turned around, took the wheel from his book case and said, "Bentele has news for you. Here's *your* wheel!"

I presented a paper on these and other vibration phenomena at a conference of the Lilienthal Gesellschaft für Luftfahrtforschung in 1943 in Berlin.

Niccolò Machiavelli (1469-1527) writes in his book, The Prince: "It may be the case that Fortune is the mistress of one half our actions, and yet leaves the control of the other half, or a little less, to ourselves." My fortune with blade vibrations consisted of the clear conditions of a turbine with partial admission. These allowed exact mathematical treatments and a solid base for their extension into all turbomachinery, including a variety of excitations and vibratory modes of static and rotating blading as well as provisions for vibration damping.

Applications of these investigations were successful with turbochargers, compressors and turbines of jet engines. The Air Ministry therefore designated me as a consultant for this field to the other jet engine companies, Junkers and BMW.

When I proudly presented my success story to American engineers interrogating me after the end of WW II, they told me in reference to Figure 2.2, "We call that a Campbell diagram." Actually, Wilfred Campbell of the General Electric Company, had investigated vibration phenomena on steam turbines. In 1924, he presented a

Fig. 2.4. Turbine wheel after test (1942).

paper "The Protection of Steam-Turbine Disk Wheels from Axial Vibration" at a meeting of The American Society of Mechanical Engineers in Cleveland, Ohio. It contains frequency/speed diagrams, in the U.S. named in his honor "Campbell diagrams" per Figure 2.2, but not diagrams for the danger level of resonances per my Figure 2.3.

I comforted myself with the fact that such accomplishments, independently made and unknown at the time, are not uncommon. May I mention as a classic example, also in the vibration field, the collapse of the Tacoma Narrows Suspension Bridge in 1940 ("Galloping Gertie"). A committee of bridge builders, investigating the disaster, encountered various controversies. Finally, Theodore von Kármán (1881-1963) convinced the bridge architects, who were thinking in static terms, that

22

the problem was a dynamic one. Prevailing winds generated Karman Vortex Streets. Their periodic shedding excited bridge structures and if their frequencies were synchronous with the bridge oscillations, the bridge would finally collapse. He proved his theory with models in a wind tunnel and established a theory and a "manual" for the design of suspension bridges.[3]

Theodore von Kármán was very proud of having found the culprit of the Tacoma Bridge collapse. He apparently did not know that, almost 100 years before, the German-American bridge builder John A. Roebling (1806-1869) had explained the failure of the Ohio River Bridge in Wheeling, West Virginia, with the same, though admittedly less scientific arguments. (The dawn of aerodynamics was still half-a-century away!) Roebling used his knowledge in establishing basic design rules for suspension bridges. His own 1867 layout led to the building of the world-famous Brooklyn Bridge, whose 100th anniversary was celebrated with grandiose style in 1983 as "The Only Bridge of Power, Life and Joy."

Full-Admission Turbines

The turbine blade vibration difficulties were only one of the imperfections of the partial-admission turbine. Turbine efficiency suffered from the alternate gas and air flow through the blade passages and the associated filling and emptying; the cooling air transport through the turbine wheel was "negative" work; also, the turbine diameter was obviously larger than that of a full-admission turbine.

With these considerations in mind, we investigated other means of turbine cooling which were internal cooling of hollow blades and cooling of the blade root. Fortunately, we were able to complement our Turbine Cooling Program with results from Government institutes, academia and industry, even from abroad.

Turbine Cooling

Analysis and design of cooled turbines included three interrelated phases, all to be carried out in my department at Hirth-Motoren: (1) heat balance, (2) mechanical design, and (3) manufacturing feasibility. We first concentrated our efforts on hollow blades and pure convection cooling with air.

Sufficient analytical methods were available for the outside heat transfer from gas to blade airfoil, for the inside heat transfer from the blade walls to the cooling air, as well as for the heat conduction across the blade cross-section and along the blade to the root and disk. Our computation sheets swarmed with coefficients and

---

[3] Von Kármán founded and organized AGARD, the Advisory Group for Aeronautical Research & Development, which had been put into effect by NATO in 1952. Shortly thereafter, I had the pleasure of meeting him in person at an AGARD Conference in Munich.

numbers named after famous discoverers such as Reynolds, Nusselt, Peclet, Prandtl, Joule, Mach, and Parsons. We used the available information in a "cook book" fashion. This was the time-consuming part, but the easy one. Converting the findings into a practical turbine wheel proved far more difficult.

The DVL Institut für Motorische Arbeitsverfahren und Thermodynamik (Institute for Engine Working Systems and Thermodynamics), headed by Dr. Fritz F.A. Schmidt (1900-1982), conducted turbine cooling investigations. They had designed and tested air-cooled turbines to their satisfaction but not to ours. Too much emphasis had been placed on the thermodynamic and cooling requirements and too little on manufacturing and operation in the field. From their experience and that of others, we became familiar with the design factors for a practical, cooled turbine blade and wheel. Also, the difficulties we had encountered with the partial-admission turbine were helpful for our considerations.

After having laid out, analyzed and designed in detail about a dozen concepts, including a variety of methods for machining and forming the blade airfoil, brazing and welding of the blades themselves and their attachment to the disk, and also having compared these with the designs of our competitors, we finally arrived at a few ground rules that satisfied the cooling requirements and the available materials and manufacturing capabilities.

Mulling over these findings, I conceived a turbine blade made entirely from sheet metal, the *Faltschaufel* (folded blade). Figure 2.5 depicts the folding process and the complete folded blade. Two reinforcement strips increase the blade strength in the critical lower blade airfoil and root sections. In addition, the inside strip provides for the proper distribution of the cooling air to the blade edges. Each individual blade is attached to the disk with an axial pin.

In our search for a competent blade manufacturer, we found two. Both liked the blade, one "as is." He calculated he could mass-produce it at a price below one Mark and still make a profit. That would have been the lowest-cost cooled-turbine blade at that time, or at any time. The second factory, the famous Württembergische Metallwarenfabrik (WMF), Geislingen/Steige, wanted to have it modified for its forming methods. We did that in cooperation with their production engineers. That became the origin of the *Topfschaufel* (tubular or bootstrap blade). It proved very successful in turbochargers and finally in the Heinkel-Hirth jet engine He S 011.

The original Topfschaufel patent application, dated November 6, 1943, names as inventors Hans Braig of WMF and me. As usual in such cases, a few of my associates were added when the blade became a reality. ("Success has many fathers, failure remains an orphan.")

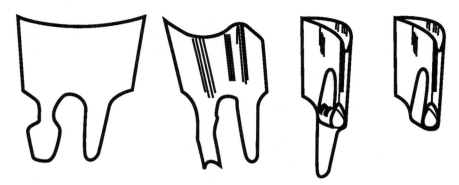

Fig. 2.5. Folded turbine blade (1943).

At a Berlin conference on cooled exhaust turbines, organized by the Air Ministry and attended by all experts in the field, opinions, suggestions and conflicts clashed head-on. Academia stood at one end of the spectrum, manufacturing at the other, and in-between the aircraft and engine people who installed the turbochargers and, most important, the Military who used them. As a Serbian proverb says, "Every gypsy praises his horse," each speaker defended his design and presented test data to prove that his was the best.

To one of my questions on blade failures, Prof. Schmidt replied, "We never broke a blade." I countered this statement, "Professor, if you baby your turbines during testing and do not establish their thermal and mechanical limitations as well as their sensitivity to vibratory excitations, you are only conducting an academic exercise and you are far away from a practical turbine. Send me one of your wheels and I'll break some of its blades." The wheel we received as a courtesy exhibited high fundamental bending frequencies of the blades which placed them in resonance with the first harmonic of the turbine stator. Running the turbine in the critical speed range produced the expected fatigue cracks and failures. Prof. Schmidt then understood my somewhat harsh comment, and we were friends again.

Apart from these practical considerations, it also became evident that the compromise to be made between aerodynamic, material and manufacturing requirements penalized the turbine efficiency. This tradeoff was different for the turbocharger and the jet engine. A turbocharger operates with a given turbine inlet temperature, whereas a jet engine improves its performance by raising this temperature. Air-cooling permits this temperature increase and thus compensates for losses associated with it.

DVL and others found different methods, beneficial in this respect, such as film or impingement cooling. However, we foresaw formidable manufacturing and operational difficulties hardly to be solved in time. We applied the same reasoning to the use of water instead of air as the cooling medium.

High-Performance Aircraft Engines

Supercharging and Turbocharging

Our cooling program resulted in satisfactory designs for air-cooled turbines. Two turbocharger applications followed this success.

For a Junkers Jumo 222, a four-row six-radial 24-cylinder 35.5-liter engine, we developed our model 9-2426 with a single-stage compressor, an airflow of 3.5 cubic meters per second and a pressure ratio of 2.4. Klöckner-Humboldt-Deutz (KHD) revived the development of its diesel engine Dz 710, a 51.5-liter flat 16-cylinder two-stroke with loop-scavenging patented by Dr. Adolf Schnürle.[4] One version was to be equipped with our turbocharger for a full manifold-pressure altitude of 15 km (50,000 feet), for which we built a two-stage compressor with an airflow of 7 cubic meters per second and a pressure ratio of 4.7.

Both compressors fulfilled their requirements on our sea-level test stand and on DVL's altitude stand at inlet air temperatures of 15°C (59°F) and -50°C (-58°F). Due to delays in the engine development, however, full-scale testing of the engine-turbocharger system could not take place.

DVL conducted an evaluation program of captured enemy aircraft and engines. Our interest focused on Rolls-Royce engines, the Merlin series. Available then were the models XLVI, XLVII and 61. The latter incorporated a two-speed two-stage supercharger with liquid-cooling of the intermediate housing and end-cooling of the charge air, all neatly packaged to the engine rear. We admired it as a marvel of engineering but were relieved when the measured Rolls-Royce aerodynamic performance data were not superior to ours. Still, the complexity of the Merlin 61 made us feel that turbocharging was the better approach to high-altitude flying. In fairness, single-engine fighters pose aerodynamic drag problems that are difficult to overcome with turbocharger installations. The Bayerische Motorenwerke (BMW) conceived and mass-produced a turbocharged engine for this purpose, the 41.8-liter 14-cylinder double-row air-cooled radial engine BMW 801 J.

The control system of our turbocharger installations presented a nagging problem in which the Air Ministry, the manufacturers of the plane, the engine and

---

[4] Schnürle became rich and famous with royalties from his loop-scavenging patent for two-stroke engines; DKW allegedly paid him one Mark per cylinder produced.

turbocharger were involved. The fingerpointing and heated arguments were temporarily relieved by one incident. A Boeing B-17 bomber, after an emergency landing in occupied France, was brought to Berlin for inspection. Of course, I was mainly interested in the Moss turbocharger arrangement. At that time, its control system was not only using mechanical but also human means, those of the flight engineer. He had scribbled on the fuselage wall for each engine power and altitude the proper numbers for turbocharger speed and manifold pressure. Pointing out this expediency to my Ministry counterpart did not bring me any Brownie points.

The Moss turbine used root-cooling of its solid blades. A detailed analysis confirmed that, due to the low heat conduction of heat-resistant alloys, this cooling effect did not reach the critical blade cross-section. Our hollow-blade program, therefore, was fully justified.

Curtiss-Wright investigated full utilization of the exhaust-gas energy. A variety of design studies, started in 1942, finally led to the Turbo Compound TC 18, a double-row 18-cylinder air-cooled engine with 3350 cubic inch displacement. In addition to a gear-driven two-speed single-stage supercharger, it incorporated three impulse turbines, each supplied with the exhaust gas from six cylinders. The turbine power was delivered directly to the engine crankshaft. This ingenious arrangement resulted in high takeoff power and low specific fuel consumption at cruising altitude and thus made possible, for the first time, nonstop flights of commercial airliners from Central Europe to the east coast of the United States.

TC 18's success started in 1950 with the model test, continued with aircraft installations, first military and then commercial, achieved still-standing long-distance records and finally included service by scores of airlines. It brought Curtiss-Wright a monopoly in this field with all associated pros and cons, e.g., some complacency and arrogance in all departments.

Curtiss-Wright got competition in developing the ultimate aircraft engine. D. Napier & Son Ltd. tried on behalf of the British Ministry of Supply to become better with a flat 12-cylinder liquid-cooled turbocompound diesel engine called Napier Nomad 2. The reciprocator simulated the Deutz-Schnürle layout; compounding consisted of an 11-stage axial-flow supercharger, driven by a three-stage axial exhaust turbine whose excess power was delivered via a Beier variable-speed drive to the crankshaft. The highly-supercharged two-stroke diesel, in conjunction with the full exhaust energy recovery, resulted in high specific power, low cruise specific fuel consumption and a competitive power/weight ratio.

Even before the project became generally known, I had heard tidbits from one of my engineers who had worked on it at Napier. Originally, the Nomad concept was even more ambitious. Nomad 1 had included a second radial compressor and two-stage axial turbine as well as a combustion chamber for boosting the exhaust gas temperature and thus turbine power for takeoff and climb. That combination apparently proved too difficult, and it was replaced with the Nomad 2 configuration. Its development progress prompted the prestigious British magazine, *The Oil Engine & Gas Turbine*, to state in its April 1953 issue:

> For the first time in the history of British aviation, oil-engined aircraft for military and commercial duties are likely to come into production in the immediate future... Now, thanks to the faith and technical ingenuity shown by the Napier company, the British aero-engine industry is in a remarkably strong position: at one stroke, the latest U.S. piston aero-engines have been rendered obsolescent, despite the long years of work that lie behind them.

Well, *one* stroke was insufficient, even the *two strokes* of the Nomad were unable to dislocate Curtiss-Wright's Turbo Compound TC 18. To the best of my knowledge, Nomad engines achieved flight tests in outboard nacelles of a Shackleton airplane but never made it to regular service operations. Finally, all high-power piston aircraft engines succumbed to the gas turbine.

May I present some considerations of Curtiss-Wright's TC 18's success and Napier Nomad's failure.

The former was the result of an evolution, starting in 1930, with a single-row radial engine. Successively were added design and material improvements, supercharging, another bank of cylinders, a two-speed supercharger drive, separate exhaust turbochargers and, finally, incorporation of the exhaust turbine into the engine package. The Nomad concept started with thermodynamics and a clean sheet of design paper with little practical experience on which to rely. The timing for this exercise was inappropriate. For aviation it was just too late, pure and simple. It may be mentioned here that the highly supercharged diesel is still alive for terrestrial applications in the form of the French Hyperbar engine and turbocompounded diesel engines for military heavy vehicles.

I wonder why the British were so proud of Whittle's invention of and success with the jet engine but, as the Nomad venture indicates, were reluctant to really believe in the jet's bright future. Unfortunately, that doubt existed not only in England but also in Germany, the country in which the world's first jet-propelled flight had taken place.

Otto-Diesel Engine

In Germany, parallel to the jet engine development, improvements in aero-engines were still being pursued, particularly with regard to their performance at high altitudes. One of the objects was a reduction of the specific fuel consumption.

Since the start of WW II, the scarcity of liquid fuels had prompted investigations into alternate fuels. One successful result was the Otto-Diesel engine. Combining the positive features of both engine types, it burns low-grade gaseous or liquid fuels that are ignited with a small amount of diesel fuel injected into the combustion chamber shortly before top dead center.

Since 1921, Professor Dr. Wilhelm Wilke (1882-1972) headed BASF's Technische Prüfstand (Engine Laboratory) Oppau, located near Ludwigshafen/Rhein. He suggested, I believe, that the Otto-Diesel principle be adapted to aircraft engines and therewith, in addition to lowering fuel consumption, eliminate alleged electrical ignition problems at high altitudes. After BASF had synthesized an "ignition fuel" with a cetane number of 300, code named R 300, the Air Ministry initiated R&D work for this combustion method, highly classified under the code name "Ring-Verfahren" (Ring combustion system).

Hirth-Motoren was chosen as an independent investigator for particular items of this project, and I was put in charge. Our test engines were single-cylinder engines of one- and three-liter displacement with carburetion or cylinder injection of the main fuel, and cylinder injection of the ignition fuel. The R engines operated like a diesel, i.e., with no airflow throttling, power control by the injected fuel, and thus leaning of the fuel/air mixture with decreasing load. We surveyed the effects of the amount of R 300 per cycle, its injection spray pattern and timing as well as those of the charge air temperatures and pressures. The typical "fishhook" fuel consumption curves for an Otto and an R engine showed the expected fuel consumption improvements as a result of lower pumping, dissociation and heat losses. Heat rejection to the coolant of a liquid-cooled three-liter R engine was 25 percent lower than that of an Otto engine and cylinder head temperatures of an air-cooled three-liter were some 25°C lower.

The Bosch and L'Orange companies supplied the two fuel injection systems, one for the main fuel and one for R 300. It provided them the opportunity to extend their technology to injections of very small amounts, on the order of a few cubic millimeters per pump stroke. Fuel pumps and injectors worked very well, but appeared rather expensive, particularly for small cylinders. Prosper L'Orange (1876-1939), one of the pioneers in diesel fuel injection equipment and founder of his company for this purpose, was aware of the technical and economic difficulties of injection systems for small diesel engines. He conceived and operated "pumpless injection" on small diesel engines that incorporated a prechamber connected to the main chamber with a Venturi tube (Figure 2.6). A mechanically operated needle valve, originally employed, was replaced by a small poppet valve that was automatically controlled by the fuel and compression pressures. I adapted the L'Orange system for R injection per Figure 2.7. During the suction stroke, the fuel pressure opens the metering valve and deposits R 300 in the cavity and the channel to the venturi. During the compression stroke, the R 300 fuel is sucked and sprayed into the small prechamber. Toward the end of compression, it ignites, resulting in a torching flame into the main combustion chamber which ignites the air/fuel mixture there. After a few adjustments, this pumpless system worked fine on one- and three-liter cylinders. With an air-cooled prechamber, a 500-hour test run was successfully completed.

The R engine suffered from one major difficulty — starting it, particularly in cold weather. We investigated the problem and became familiar with the intricacies of diesel starting and the differences between liquid- and air-cooled engines. In tests with three-cylinder and complete engines, we established the optima for the amount of R 300 and its injection timing, and additional means such as throttling and preheating the charge air. Finally, we achieved satisfactory starts at acceptable starting speeds, but the starting problem remained the "Achilles heel" of R engines. In combination with the advent of jet propulsion, it probably contributed heavily to its abandonment at the time.

Before that, our Hirth R program came to an unexpected, sudden stop. I was quietly scrutinizing a test report in my office when one of my mechanics shouted through the open window, "Our building 24 exploded!" In disbelief, I bicycled to the site and saw the catastrophe. Half of the building lay in ruins, and one technician was seriously burned in the face and hands and rushed to the hospital. There was

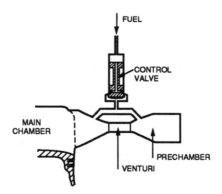

Fig. 2.6. L'Orange pumpless injection (schematic).

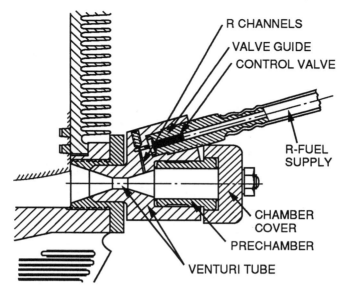

R CHANNELS
VALVE GUIDE
CONTROL VALVE

R-FUEL
SUPPLY

CHAMBER
COVER
PRECHAMBER

VENTURI TUBE

Fig. 2.7. Pumpless R injection (1942).

no fire, though. The usual first thought of sabotage proved wrong. A careful investigation resulted in the following sequence of events. One part of the building, where the explosion occurred, consisted of eight engine test cubicles, each with a control room, and a basement underneath the test cell. In one of these rooms, a gasoline line had sprung a minor leak dripping fuel to the floor. Unfortunately, the door to this room was closed so that the ensuing fuel/air mixture eventually reached the proper ratio for an explosion. Ignition was probably triggered by fan blades rubbing on the housing. This experience taught me a lesson: Should, in an enclosed space, a combustible fuel/air mixture build up, a spark for its ignition will always come from somewhere.

In the mid-1960s, Julius E. Witzky (1903-1981) of the Southwest Research Institute reinvented L'Orange's pumpless injection, apparently only aware of one publication with a mechanically operated needle valve. The Porsche Company used a prechamber similar to my pumpless R system per Figure 2.7 for the ignition of lean mixtures. I myself suggested the pumpless prechamber system for this purpose. Lean burning, however, was unable to fulfill stringent exhaust emissions requirements. The Peugeot Company diligently pursued this approach, but had to abandon it; the new European exhaust emission standards had posed an insurmountable hurdle.

31

A new scientific truth does not usually assert itself by convincing its opponents and by their enlightenment but because its opponents gradually die and the oncoming generation is from the beginning made familiar with the truth.

Max Planck (1858-1947)

Chapter 3

The Dawn of the Jet Age

In the late 1920s, aviation was at a crossroads. The old guard tried rather successfully to push aircraft with sheer force to faster speeds and higher altitudes. The ecstacy for speed was exemplified by the international quests for the Schneider speed-race Trophies. A new school sensed that the reciprocating engine/propeller drive for aircraft propulsion approached barriers, some of which were regarded as insurmountable with orthodox means. This realization was particularly true for increasing aircraft speed to the speed of sound and beyond.

Thinkers and tinkerers, independent of each other, worked toward solutions. All kinds of propulsion were "in the air," such as ramjets, turbojets, engine/fan jets, and rockets. Occasional public demonstrations illustrated the hidden turmoil brewing in aviation, for example, Fritz von Opel's (1899-1971) drive of a rocket-propelled automobile and his rocket-assisted glider flight. A variety of investigators and institutions were busy in this new and fascinating field, some on paper only and some experimenting with hardware.

From Turbochargers to Jet Engines

We at the Hirth Motoren Company were in the process of developing and manufacturing turbochargers for large reciprocating aircraft engines, but we were too small and financially weak to enter an entirely new field. We were even considered too small to stay competitive on our own and, therefore, were put up for sale to the highest bidder.

Representatives from all major aviation companies visited and inspected us, Daimler-Benz, Junkers, BMW among others from the engine field, Messerschmitt, Heinkel, Dornier, and smaller ones from the airframe companies. Our location in a forest at the outskirts of Zuffenhausen, a suburb of the state capital Stuttgart, and our physical outlay were unique, with separate buildings for the Applied Research, Design and Test Departments, Materials Laboratory, Experimental Machine Shop, Component and Full-Scale Test Facilities, Administration, Cafeteria, Dining Room, Indoor Gymnasium and Outdoor Sports Arena — no swimming pool, though. One of the suitors — I believe Dr. Claude Dornier (1884-1969) — was not impressed and declined to bid, commenting, "I want to spend my money on a factory, not a health spa." Twenty years later I found an Industrial Research Park in Tuxedo, NY, with the identical arrangement, except for the missing camouflage netting we then had over the roads interconnecting the buildings.

Ernst Heinkel (1888-1958), the famous aircraft designer and manufacturer, had the inside track and bought us lock, stock and barrel in April 1941. Putting bits and pieces from official statements and from the grapevine together, we realized that the offering and inspection of the other companies had been a pro forma charade. The Air Ministry had promised Heinkel his acquisition of Hirth Motoren on the condition of a successful flight demonstration of his twin-jet fighter, the He 280. The Heinkel organization was allegedly lacking the expertise and facilities to fully develop the fighter powerplant, the jet engine He S 8 A, and to manufacture it. We of Hirth Motoren offered the potential to eliminate both shortcomings, which we did — and more.

It was the first takeover I experienced. At the time, there were no textbooks yet on successful strategies both for individuals and for the acquired entity. We put our own G 2 intelligence network in operation to be prepared for facts and imponderables on Professor Dr. Ernst Heinkel and his jet engine venturers. As a close-knit group, confident of our general and professional capabilities, we faced the challenge squarely.

I was obliged now to wear two hats for my R&D responsibilities: the existing one for reciprocating engines and turbochargers and a second one for jet engine components. The gradual familiarization with the new field soon received a severe shock. On a June morning, loud Wagnerian march music from a tenant's radio awoke us: it was the theme for the announcement of Hitler's attack on the Soviet Union. Immediately to mind came Germany's two battle fronts in the First World War, Napoleon's Russian campaign of 1812, and Japan's invasion of the Asian mainland. The war would be with us for a long, long time, with no negotiated settlement in sight. Though the Russian battlefront exerted no direct impact on our work at the plant, our outlook on life became a very simple one — survival.

Pioneer Jet Engines, He S 1 to He S 8 A

Ernst Heinkel himself was well known as an energetic and flamboyant entrepreneur and a rugged individual. From his youth he was fascinated with flight and wanted to be "first" in aviation. In the beginning he tried it alone; later, he attracted good engineers and scientists, often the best. He succeeded in becoming number one and put his name into the history books. In 1938, he was awarded, together with his archrival Willy Messerschmitt (1898-1970, fighter planes) and Ferdinand Porsche (1875-1951, Volkswagen), the German National Prize for Art and Science. For this honor he received, among others, a congratulatory letter from the American Embassy.[1]

Heinkel, born in Grunbach of solid Swabian stock, had received his engineering education in Stuttgart, but left his native south for the north of Germany before World War I. He returned south only for short visits on special occasions, for example, in 1925 to pick up an honorary Doctor of Engineering degree from his alma mater. Though Hellmuth Hirth, our late founder, had been his lifelong friend, Heinkel's attitude toward us Swabians, mostly regarded as square, was unknown and of some concern.

One consolation in this respect was the trouble with his four-engine, two-propeller He 177 heavy bomber. The Air Ministry arranged frequent meetings with him, which we hoped would leave him little time to meddle in our work.[2]

On August 27, 1939, Heinkel had achieved another historic milestone with his He 178, an experimental airplane that demonstrated the first purely jet-propelled flight and ushered in "The Jet Age." This airplane's engine, the He S 3 B, had been designed and built by a small team headed by a young physicist carrying the awe-inspiring name Dr. phil. Hans-Joachim Pabst von Ohain (1911-).[3]

He was one of the thinkers mentioned above. While working on his doctoral thesis in physics at the University of Göttingen in late 1933, he started to dream about new propulsion systems, settled for a simple turbojet and tried to prove his theories with a rudimentary working model, a back-to-back radial compressor-turbine rotor. Max Hahn (1904-1961), the master mechanic of a local garage, constructed it for him at a cost of 1000 Marks, equivalent to $240 (see Figure 3.1).

---

[1] For his aviation achievements, Heinkel was inducted in 1982 into the International Aerospace Hall of Fame (IAHF) at San Diego.

[2] In the mid-1960s, Roger Lewis (1912-1987), President of the General Dynamics Corporation, experienced similar controversies with the Pentagon on the F-111 supersonic bomber. Secretary of Defense Robert S. McNamara (1916-) scheduled meetings with him for every Saturday and aggressively opened these with words like, "Let's talk about your bad airplane." Not having been privy to the proceedings of either of those highest-level conferences, I cannot pass judgment on their contributions to solutions to the problems. In my own experience, similar meetings delayed rather than promoted the development process toward a satisfactory product. Reporters of the 1980s summit meetings apparently agree with my opinion as indicated by their phrase, "Heads of State don't do windows."

[3] Emigrating to the U.S. in 1947 for the Wright-Patterson Air Base in Dayton, Ohio, he shortened his name to Dr. Hans-Joachim von Ohain, and finally to Dr. Hans von Ohain or Hans von Ohain, Ph.D., much to the relief of encyclopedia and other publishers.

Fig. 3.1. Max Hahn with working model of a jet engine (1935).

Though self-sustaining speed proved elusive, the model tests indicated (1) the soundness of the principle of the turbomachinery and (2) the need for further development efforts, particularly on the combustion system. He felt the expenditure needed would exceed his private means. In retrospect, that was intuitively a correct appraisal. Combustion alone required major development efforts over the next four decades, and still does. In 1972, the whole world was made aware of unresolved problems in a dramatic demonstration: In the departure of President Nixon from Shanghai's airport shown on TV, the Air Force One could hardly be seen for the smoke emanating from its engines.

Professor Robert W. Pohl (1866–1961), the mentor of his doctoral program, came to Dr. von Ohain's rescue. He introduced him to Ernst Heinkel, who hired him on the spot to be in charge of a Sonderentwicklung Department (Special Development, in Lockheed's jargon called "Skunk Works"). Heinkel insisted on keeping that unit as his own private enterprise, immune from any government interference and without any government knowledge or financial support.

Having established the separate department and assembled its small team (of course, including Hahn), Heinkel was extremely anxious to fly a jet-propelled

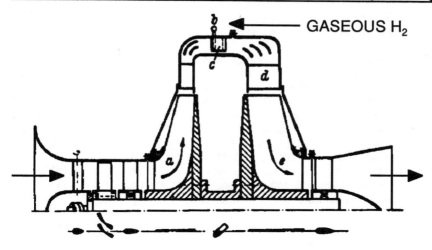

GASEOUS H₂

Fig. 3.2. Radial turbojet He S 1 (1936-37).

airplane as soon as possible. He gave Dr. von Ohain a target of 600 kilopond (1325 lb) thrust for a suitable flight engine to be ground-tested in about a year's time. That was, to put it mildly, a "tall order."

Aware of the skimpy data of the model tests and the combustion difficulties, Dr. von Ohain decided to take two steps toward a flight engine: (1) to design a minimum risk engine to demonstrate the jet principle in a convincing and impressive manner, and (2) to develop simultaneously a suitable combustor. The demonstrator engine consisted of the proven back-to-back radial compressor-turbine configuration with the addition of an axial-flow entrance stage. To eliminate any combustion problems, the engine was to be fueled with gaseous hydrogen. Figure 3.2 depicts the cross-section of this He S 1 (He for Heinkel, S for Strahltriebwerk = jet engine). It was built in late 1936 and tested in early April 1937. Producing a net thrust of 250 lb at 10,000 rpm, the He S 1 met all performance and handling expectations, the latter due to the low moment of inertia of the rotor and the stability of the hydrogen combustion over the full operating range, including acceleration and deceleration. The psychological effect was even more impressive: Heinkel and his airframe engineers were now fully convinced of the true potential of jet propulsion.

The development of liquid-fuel combustion took a fairly long time; it led to a wrap-around combustor and, after some modifications finally to the flight engine

He S 3 B (Figure 3.3), which powered the He 178 (Figure 3.4) on the world's first jet-propelled airplane flight in Marienehe on August 27, 1939.[4]

The impact of Heinkel's pioneering jet engine development, which culminated in this flight, was momentous and accompanied by some ironies. Various individuals and groups had worked on jet propulsion. However, the old established engine companies failed to perceive the tremendous potential of this field and stubbornly refused to pursue it, except as a token activity. As an example, during negotiations with the Air Ministry, Professor Otto Mader (1880-1944), the venerable technical director of the Junkers Aircraft Engine Division, had uttered the words, "The nonsense of jet engines should never come into this company!" In the new environment, the Air Ministry would now offer jet engine development contracts on silver platters, and they were gladly accepted by Junkers, BMW, Daimler-Benz and by other smaller institutions.

In 1938, the U.S. Navy had ordered an investigation of jet propulsion for airplanes. The Committee of the National Academy of Science stated:

> In its present state, and even considering the improvements possible in adopting the higher temperatures proposed for the immediate future, the gas turbine could hardly be considered a feasible application to airplanes, mainly because of the difficulty in complying with the stringent weight requirements imposed by aeronautics.[5]

Having paid with his own money for the jet engine development and flight demonstration, Heinkel relished these accomplishments and was ready to work with the government now. He accepted a contract for a twin-jet fighter plane. The von Ohain team developed and built its engine, the He S 8 A (see Figure 3.5). The successful flight of the world's first jet fighter, the He 280 on April 2, 1941, brought Hirth Motoren — and myself — into jet propulsion. I am hardly ashamed of having entered the jet palace through the back door; the road from turbochargers to jet engines can be considered logical and natural. In turbojets, hot gases are generated in a combustor; turbochargers obtain them from an engine, that's all. Furthermore,

---

[4] Replicas of the He S 3 B engine are on permanent exhibit in the National Air & Space Museum in Washington, D.C., and the Deutsches Museum in Munich.

[5] Noteworthy items to this report: It was issued at a time when a turbojet was already running in Heinkel's Skunk Works that demonstrated its four-fold advantage in power-weight ratio over a conventional recip/propeller system. In England, Frank Whittle (1907-) had demonstrated the principle with a crude experimental jet engine in April 1937; subsequently, he concentrated on a flight engine. Theodore von Kármán, a member of this committee, learned this lesson: "If you accept membership on a board, you should attend its meetings and read its reports." He was away in Japan at the time the report was written and had allowed his name to be used without reading it. "I was furious at myself," he said after having learned what had happened. In 1941, the U.S. recovered from this ominous setback with remarkable speed due to the alertness and initiative of General Henry A. (Hap) Arnold (1886-1950) of the Army Air Corps.

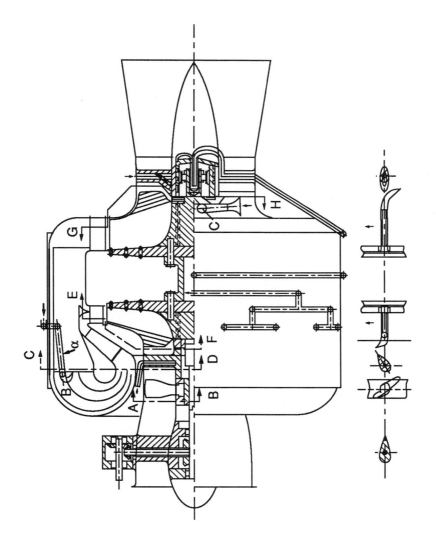

Fig. 3.3. He S 3 B engine which powered world's first jet-propelled flight (1939).

Fig. 3.4. He 178, world's first jet plane (1939).

40

other pioneers traveled the identical route; prime examples are Dr. Anselm Franz (1900-) of the Junkers, and the General Electric Company through Dr. Sanford A. Moss (1872-1946).

Axial-Flow Jet Engines

Wagner-Müller-Heinkel; He S 30

Ernst Heinkel planned to transfer all jet engine developments from the Baltic seashore to his newly acquired Stuttgart engine plant. It proceeded in stages according to the availability of new buildings and test facilities. The first team arriving in Stuttgart was that of Max Adolf Müller (1901-1962) with the jet engine He S 30. It represented the most advanced aerodynamic design with axial-flow turbomachines (see Figure 3.6). The compressor produced a pressure ratio of 3.2 in only five stages due to its blading with 50/50 percent reaction in the rotor and stator (so-called symmetrical blading). The engine layout was way ahead of the competition.

We newcomers were told that jet engine design and development work consisted of 80 percent science and theory and only 20 percent experimentation, whereas in our activities the ratio was allegedly the reverse. At first sight we believed this and were impressed, but gradually we discovered evidence to the contrary. Examples were a variety of turbine designs with shortened and with elongated blading, a two-stage configuration and variable turbine inlet guide vanes.[6] Also, the circuitous history of the engine put a damper on our enthusiasm for gas turbine theory and practice.[7]

Some gas turbine proponents still claim today that they are able to design turbomachinery powerplants close to the given performance specifications and need test-stand operation only for verification of the design data. It may even be true for large turbomachines but is hardly so for medium and small power outputs.

Germany's first axial-flow jet engine originated with Herbert Wagner (1900-1982), a professor of aeronautics in Berlin who had started investigations on jet propulsion with axial-flow machines in 1934. In due course, Dr. Rudolf Friedrich (1909-) designed the 50/50 compressor blading and Max Adolf Müller pursued development of the jet engine at the Junkers Magdeburg plant. In 1939, when Hans

---

[6] It took courage — or naivety — to put controlled variable vanes into the hottest engine section. Nowadays, they are used for the work turbine of vehicular powerplants.

[7] As a corollary, in the fifty years of my professional career, I had on some occasions to work with an engineer where I was ambivalent whether he was a genius or a fake or alternated between the two extremes. Nature often made the decision for me when one of his grandiose design concepts failed miserably in practice. Sir Rudolf Bing (1902-), General Manager of the Metropolitan Opera House in Lincoln Center, New York City, coined a remark about his performers which applies equally well to engineers: "It is much worse to be a mediocre artist than to be a mediocre post office clerk."

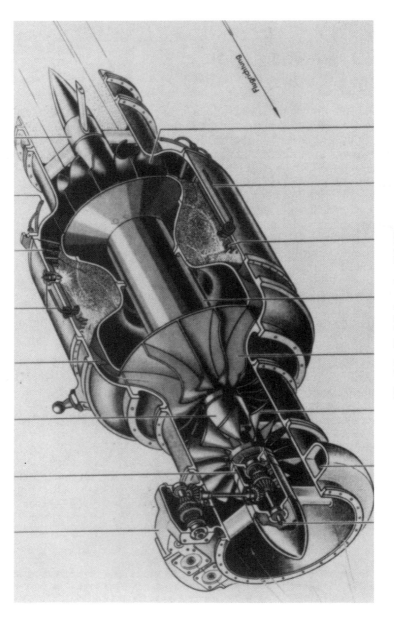

Fig. 3.5. He S 8 A engine (1941).

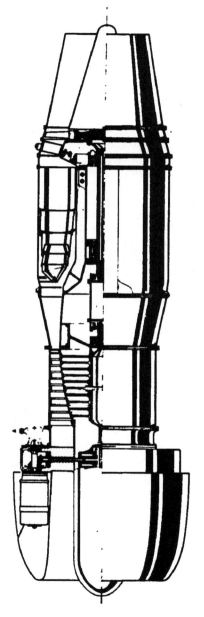

Fig. 3.6. He S 30 jet engine (1942).

Mauch (1907-1984) and Helmut Shelp (1911-) of the Air Ministry Research Division initiated a broad jet engine development program with the aircraft engine companies, Dr. Anselm Franz (1900-) of the Junkers Engine Division in Dessau refused to base a "production" engine on this design but rather chose a less risky configuration. Müller then transferred to Heinkel to continue the engine's development under the new designation He S 30.

He frowned on our assistance in solving the engine's many problems. At that time, another blow hit me. Riding in a tramcar to work I noticed the bold headline in another passenger's newspaper, "Schüsse im Pazifik" (shots in the Pacific), referring to Japan's air attack of Pearl Harbor. With the resulting entry of America into the war in Europe, the question would then definitely not be if Germany would have to surrender but when. That time was, however, hardly considered imminent and my work continued unaffected for the time being.

In the Fall of 1942, the He S 30 development had progressed to the stage where all performance objectives were met. Efficiencies, fuel consumption, and particularly thrust per engine weight and per frontal area were superior to contemporary engines. The He S 30 had paved the way for advanced axial-flow jet engine technology with features that are still in evidence today. However, the thrust was marginal for the contemplated aircraft; the competitive engines were closer to the production requirements and the project was therefore abandoned. The He S 30 project proved the old adage: "For success in the marketplace, the right product has to be available at the right time."

Junkers Jumo 004

Following Heinkel's jet engine demonstrations, the Junkers Company had received a government development contract for an axial-flow jet engine of the 2000-pound thrust class at a time when the technology for high-power concentration was still in its infancy. The Wagner-Müller engine, for example, was not yet able to demonstrate its potential on the test stand. In view of a desirable early production date, Franz selected engine components that were not unnecessarily saddled with basic unresolved problems.

The cross-section of the basic engine (designation Jumo 004) indicates the following configuration: An eight-stage compressor, a six-can combustor, a single-stage axial turbine, a jet nozzle with a movable inner cone, and a control system for fuel flow and jet nozzle area. This cautious conservative approach paid off: The 004 A engine was first run on the test stand in October 1940, one year after go-ahead.

44

The compressor was based on experimental results with superchargers developed in cooperation with the Aerodynamische Versuchsanstalt in Göttingen (AVA) (Aerodynamic Research Institute).[8] It utilized nonsymmetric 100 percent reaction blading, i.e., pressure rise by the rotor blades only. The superiority of symmetric blading is demonstrated by the following comparison: The symmetric He S 30 compressor finally achieved a pressure ratio of 3.2 with five stages; the nonsymmetric Jumo 004 needed eight stages for its highest pressure ratio of 3.14.

The sheet-metal stator vanes, originally cantilevered from the outside, suffered from vibration difficulties. My Air Ministry status as a consulting engineer for vibration problems offered me the opportunity to familiarize myself with the 004 engine and to assist in ameliorating its vibration difficulties.

The development progressed satisfactorily. Two 004 A engines powered the Messerschmitt twin-jet fighter aircraft Me 262 on its first flight on July 18, 1942, which was over one year after the first flight of Heinkel's He 280. This successful flight prompted the manufacture of 80 more engines for testing on the stand, in the aircraft and for general evaluation. The 004 A design was, however, unsuitable for mass production due to its excessive weight and large content of strategic materials with ingredients such as nickel, cobalt and molybdenum.

The redesigned version, designated 004 B, incorporated an improved compressor and operated at a slightly higher turbine inlet temperature. It fulfilled the thrust and weight requirements and was available in June 1943. From the beginning, it experienced serious turbine rotor blade failures, resulting in engine malfunctions and aircraft crashes. The Junkers team tried diligently to resolve the problem with analytical and experimental investigations. The Air Ministry, impatient with the progress made and under severe pressure from the generals at the front, scheduled for December 1943 a conference to be held at the Junkers Dessau plant and to be attended by experts from government, academia and industry associated with turbine design, manufacturing and operation. Since I was the lowest of this distinguished group in both age and salary scale, I kept my powder dry, at first only listening to the arguments put on the table. The lively discussion related almost exclusively to areas such as material defects, grain size, surface quality and roughness, manufacturing tolerances and their deviations; in my opinion, secondary effects at best. Based on my experience with DVL-Hirth turbochargers, I felt the conference was ready to hear from me. I stated to the eager audience that I was convinced of having a definite solution to the blade problem, and an expeditious one

---

[8] AVA Göttingen enjoyed an excellent worldwide reputation based on its venerable members. Those related to gas turbines and jet engines included: Professor Dr. Ludwig Prandtl (1875-1953), father of modern aero- and hydrodynamics, honored with the Prandtl number; Professor Albert Betz (1885-1968), aerodynamics, initially mainly propellers; Walter Encke (1898-1984), axial-flow compressors; and Ludolf Ritz (1908-), rotary regenerators. Before the First World War, Theodore von Kármán worked in Göttingen with Prandtl. At that time he proved with pure mathematics the shedding of vortices which made him famous with his Kármán Vortex Street.

at that. The following design features, the six combustion cans before and the three struts of the jet nozzle housing after the turbine, are the culprits. They induce forced excitation of the turbine rotor blades; its sixth order is in resonance with the blade natural bending frequency in the upper speed range. The predominance of the sixth order is the result of the six combustor cans, undisturbed by the 36 turbine nozzles, and the second harmonic of the three struts downstream of the turbine rotor.

In the 004 A, this resonance fortunately was above the operating range; in the 004 B it had slipped into it for two reasons, namely the slightly higher maximum engine speed and the lower blade natural frequency due to the higher turbine temperatures. My analysis was convincing in both logic and elegance. Elimination of these turbine blade fatigue failures was extremely simple: raising the blade natural frequency by initially shortening the blade and finally by increasing the airfoil taper and by slightly lowering the engine maximum speed. My recommendations, immediately put into effect without interruption of production, were 100 percent successful, and everybody was happy.

My own happiness over this achievement was damped by the increased vulnerability of my family to air raids, now also carried out in daylight by American B-17 bombers. We were the victim of the long-term strategy of William S. Knudsen (1879-1948), i.e., the mass-production of engines and aircraft with new tools and new factories. A captured Moss supercharger with a housing redesigned from a casting to a stamping already had indicated to us this transition. Therefore, we decided in the Spring of 1944 to evacuate to Magda's parents house in Heilbronn, 25 miles north of Stuttgart, a presumably less attractive air-raid target. This move proved to be prudent in view of the Allied invasion in June, and shortly afterward the fiery destruction of the Stuttgart apartment building we had lived in.

Our third daughter, Brigitte Regina, was born during an unannounced air raid on Heilbronn. It was thus time to look for a place in the country. Hubert, a friend of mine, located for our two families a suitable one in Brachbach, a hamlet 40 miles northeast of Stuttgart. It consisted of half a dozen small farms, had no factory, no tall chimney and no church spire, no river or brook, no major road and, of course, no railroad. In short, it was the ideal place for a smooth transition from war to peace. That's what we and everybody else thought. My family found accommodation in a country inn and farm which was operated by a widow, her daughter and a farm worker from Poland. Since the beginning of the war, the inn and restaurant had been closed and were made partially available to us. I had rented a room in the factory's neighborhood from where I commuted to work on foot or on a bicycle. For weekend visits to my family, I could take a train to a station five miles from Brachbach, and take a bus from there; sometimes I used my bicycle.

My professional cooperation with Junkers extended further with our 15,000-horsepower compressor test facility in Dresden which they were able to use for 004 compressor investigations under various inlet and pressure level conditions.

The Jumo 004, the world's first mass-produced jet engine, powered the Messerschmitt Me 262 twin-jet fighter, the Arado Ar 234 reconnaissance aircraft, and the experimental Junkers Ju 287 four-jet bomber.

In 1944, the Me 262 flew a record 1004 km/hr (624 mph); its operational speed of 850 km/hr (530 mph) made it a superior fighter aircraft. Approximately 6000 Jumo 004 engines and 1400 Me 262 aircraft were produced. In surveillance flights over England the Arado achieved altitudes of 14 km (46,000 feet).

In 1944-45, I was an occasional eyewitness to the glories and calamities of jet aircraft operations with the 004-powered Me 262. After a fierce daylight raid on the Daimler-Benz plant at Stuttgart-Untertürkheim, I returned with my colleagues from our air-raid shelter, a huge natural cave near our workplace. We noticed two specks high up in the blue sky, then suddenly three: one falling in spirals and exploding on impact, one disappearing fast on the horizon, and one, a parachute, floating slowly down to earth. Its Allied pilot described the event as follows: "After having taken pictures of the damaged plant, I was happily heading home. Unprepared and without any warning signs, I heard a few shots, experienced loss of engine power and aircraft control and was forced to eject. I had never observed the jet fighter before or after I lost my aircraft."

The military airfield of Schwäbisch Hall, a few miles from Brachbach, served to train pilots and ground crews with the Me 262. On visits to my loved ones, by train or bicycle, I was within this operational area. In the words of a general, "Angels were pushing jet aircraft through the air," the smooth flights and distinctive jet whine were pleasant to observe. However, on some occasions the engines experienced flameouts during landing and overheating at takeoff. Benign emergency landings — one in our backyard — and fiery aircraft crashes also happened. To me the early jet age presented a mixture of exaltations, tribulations and hardships and, for us engineers, the realization of a lot of work still to be done.

BMW 003

As briefly described above, axial-flow compressors may generate the pressure rise in the rotor only (as in Jumo 004) or in both rotor and stator (as in He S 30). The engineer Helmut Weinrich went a step further and suggested two counterrotating rotors which would produce high pressure ratios with the shortest axial length. The BMW Company had conducted aerodynamic and design studies on the Weinrich concept and found it correct, but associated with severe penalties in development and manufacturing.

With this background, BMW selected a single-shaft configuration for a jet engine of 1700 to 2000 lb thrust, to be developed under government contract. Dr. Hermann Oestrich (1903-1973) was in charge of this project at the Bramo plant in Berlin-Spandau which BMW had acquired from the Siemens Group.

47

This engine, designated BMW 003, shows vague similarities to the Jumo 004. Some of its features were more advanced: the seven-stage compressor (pressure ratio 3.1), annular combustor, and air-cooled turbine whose blade design was based on turbocharger experience but with a somewhat queer attachment to the disk. The latter was plagued with fatigue cracks and failures. In my discussions on the subject, I made recommendations for remedial changes by lowering the excitation forces. I was, however, rebuffed with the claim these had already been tried without success. First I left it at that. Then I learned from my friends in our machine shop, which fabricated 003 engine components under subcontract, that a new change order incorporated my suggestions. This turn of events made me think again about my visit to Spandau—the last one.

BMW 003 engines also went into production and operation but in much smaller numbers than the Jumo 004.

From the He S 8 to the Second-Generation He S 011

He S 8 A

In contrast to the He S 30 project, participation of Hirth engineers in the development of the He S 8 to the production stage was most welcome and appreciated. The North and South teams soon established a close relationship as well as a frank rapport with von Ohain. Technically he became my immediate superior though I never had the feeling that he acted as my boss. I hardly needed one, either. In general, too, he favorably differed from the Prussian Junker image, so much disdained by Swabians.

Our plant, officially called "Ernst Heinkel A.G. - Werk Hirth Motoren," or "Heinkel-Hirth" for short, retained the old management under the direction of Curt Schif. Ernst Heinkel preferred a more personal link and established that permanently with Harald Wolff (1880–1969), sent from Berlin and put in charge of jet engine development. The individual authorities of Schif and Wolff were spelled out on paper, but left some ambiguities in practice. We made the best of this duality or, in other words, we occasionally took advantage of it.[9] From this experience I often wondered whether a so-called "matrix" organization really leads to maximum results.

Rapid development progress reflected the harmony in the engineering trenches. In July 1942, an He S 8 A was run with a remarkable continuous thrust level. Our delight was of short duration, being interrupted by the first flight of the Messerschmitt

---

[9] A typical example comes to my mind. I definitely wanted to attend a conference in Berlin. I knew pretty well who of the two would dislike my trip, for one reason or another. So I asked the other one for approval with the remark, "Your colleague is, of course, against my traveling to Berlin." His answer always was, "Please go!" So I did and took Magda along to visit with our old friends.

fighter powered by two Jumo engines. That flight resulted in a preproduction order of 80 engines and subsequently in the selection of the Me 262/Jumo 004 combination for production.

After years of pioneering efforts and successes, Heinkel had lost the jet fighter competition and the chance for jet engine production for technological and also political reasons. Messerschmitt enjoyed predominance and vast experience with his piston engine-powered Me 109 fighter, whereas Heinkel's forte was in the commercial airliner, medium and heavy bomber field. Furthermore, Junkers was known as a capable and reliable engine manufacturer and already had worked on the 004 production aspects for a long time. Heinkel-Hirth's He S 8 A performance was at that point superior in thrust/weight ratio, overall length and fuel consumption and only slightly inferior in thrust per frontal area. Its severest shortcoming was, in my opinion, its radial turbine, for a number of reasons.

The material temperatures, especially at the high-stress regions, were lower than those of axial rotor blades. The overall weight requirement for strategic materials was, however, larger. The radial blades — we nicknamed them "elephant ears" — were difficult to attach to the hub and also suffered from fatigue failures. An air cooling scheme was remote, if feasible at all, i.e., thrust growth by raising the turbine inlet temperature was hardly possible. Last, but not least, production experience for this radial turbine size was virtually nonexistent. Conversely, the axial turbine was able to benefit from the vast field of steam turbine design, manufacturing and operation. Junkers, for example, cooperated closely with the Allgemeine Elektrizitäts-Gesellschaft (AEG). Borrowing from steam turbine technology was helpful for the mechanical design and production of gas turbine blades but, unfortunately, not for the best aerodynamic layout of the blade airfoil profiles. A design tailored to the higher expansion ratio of a gas turbine would have offered higher efficiencies than those obtainable with steam turbine practice.

In view of the pros and cons of the radial-inflow turbine, its use was for years restricted to low gas-flow rates; small turbochargers employ them exclusively nowadays. In 1969, a quarter of a century after the He S 8 jet engine, the Norwegian A/S Vapenfabrikk Kongsberg produced with its KG 2/3 a radial turbine of approximately the He S 8 size for industrial and marine use, therewith (in their words) "marking the beginning of a new era for gas turbines." Kongsberg used superalloys for the turbine, a one-piece forging of Nimonic 90 for the impeller and a one-piece casting of 713C for the exducer, i.e., materials for us unavailable and "verboten."

In the late 1970s, development efforts for an automobile gas turbine included air-cooled radial turbines. These were unsuccessful and then replaced with ceramic materials. It is noteworthy that the Garrett automobile gas turbine design simulates the back-to-back radial compressor/turbine concept of von Ohain's early jet engines.

In early 1940, the selection of a powerplant for the He 280 had posed a dilemma. Time was of the essence. The airplane required a small-diameter engine for under-the-wing mounting. It was highly unlikely that the all-axial Wagner-Müller engine would be ready, and the He S 3 B concept suffered from an intolerably large diameter. To save time, experience with radial turbomachinery was utilized, and the He S 8 design evolved, hopefully posing no new design and performance problems. This shortcut provided a handsome profit for Heinkel: It enabled him to acquire Hirth Motoren with its excellent manufacturing and test facilities as well as its pool of outstanding engineers and scientists. The He 280/He S 8 A project was, however, no exception to the rule that expediency seldom pays off in the long run. Heinkel's opportunity to enter the jet fighter and the jet engine production business was lost for the time being.

Simultaneously, the Heinkel team started basic compressor studies on concepts that combined the best of radial- and axial-flow machines; they provided noteworthy insights into advanced jet engine compressor technology.

He S 011 Development

One result of the fundamental compressor studies was a *Kombinationsverdichter* (combination compressor), its cross-section and blading depicted in Figure 3.7. This compressor includes a front axial stage, a *Diagonalrad* (mixed-flow wheel), followed by three axial stages with symmetric blading. All stages operate at approximately the same subsonic tip Mach number. Its aerodynamic advancements were a high overall pressure ratio with high loading of all stages and high stage efficiencies, a flat characteristic for good matching at the changing operating conditions of takeoff, altitude, maximum flight speed and snap accelerations and decelerations. It led to a very simple engine control system, eliminating the continuous control of the jet nozzle area. Under full-power conditions, control of the fuel flow to full engine speed provided, with one constant jet exit area, equal turbine inlet temperature for all flight speeds at sea level and all altitudes. For engine starting and a higher idle speed for landing, a larger exit area was found necessary. Apart from the resulting two-position jet nozzle, the engine contained no variable geometry or bleed-off.

Mechanically, the compressor envelope provided space for accessories without enlarging the frontal area over that of the last axial stage. Its compactness also permitted a single engine rotor, supported by only two bearings, and thereby a thrust balance that simplified the front bearing design.

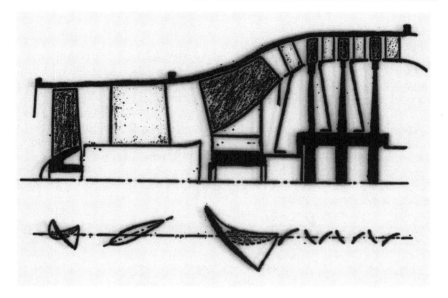

Fig. 3.7. Cross-section and blading of He S 011 compressor (1942).

Mainly on the basis of the favorable results of these investigations, the Air Ministry in September 1942 granted Heinkel a development contract for a second-generation jet engine (designation He S 011), offering a quantum step in power concentration and performance as follows:

Maximum thrust, static sea level 1300 kilopond (2866 lb, 12.75 kN) with growth to 1500 kp (3307 lb, 14.7 kN); weight under 900 kg (1985 lb); pressure ratio 4.2; altitude capability 15 km (50,000 feet); fuel consumption below 1.4 kg/kp.h (1.4 lb/lb-hr).

Figure 3.8 depicts schematically the engine layout.

I was put in charge of engine components and assembled a design, test and development team composed of my turbocharger experts at Hirth Motoren, remnants of the Müller jet engine group (he himself had left earlier), and an uneven mixture of newcomers.

Compressor

For compressor investigations, we were well equipped with our own 1600 kW electric test stand in Zuffenhausen and a steam-turbine-driven 15,000 hp stand in Dresden at Brückner, Canis & Co. (where Friedrich was now in charge). Also available for leasing were a Diesel engine stand at MAN in Augsburg and an electric 2800 kW drive at AVA Göttingen, under Encke's jurisdiction.

Aero/thermodynamic data were measured at all stages of the mixed-flow and the full compressor. The acquisition system at the time was mechanical and manual: tubes with three holes for total pressure and flow direction at various depths of the flow path; U-tubes filled with water or mercury and precision manometers; Betz micromanometer for mass flow, and mercury thermometers enclosed in a ram tube for total temperatures. Determination of the adiabatic efficiencies was based mainly on temperature rise. Additionally, the power input to the compressor was measured as torque, in Augsburg by means of a pendulum support of the compressor, and in the other cases by torsion drive shafts.

Data analysis and conversion of the raw data into performance maps was conducted by two "personal computers": a former seamstress equipped with slide rules, logarithmic and other tables, a sharp pencil and a handful of preprinted forms, and an art student for the graphics. It took only slightly longer to have performance diagrams of the tested compressor on my desk than it did twenty years later with sophisticated instrumentation and computerized equipment.

Testing started in May 1943, nine months after go-ahead.

The mixed-flow wheels required a number of modifications. The expected performance occurred at the inner third of the flow channel, but deteriorated outward. We postponed improvements because the overall performance, mass flow, pressure ratio and efficiency were considered adequate for initial engine tests. Even then, I felt sorry for the axial stages which had to deal with the rather awkward flow profiles from the mixed-flow wheels.

Rather than arrive at the optimum configuration of the axial stages analytically, we employed adjustable stators. A variety of settings was explored on the test stand, once three settings in one night. This experimental iteration led to a satisfactory performance of the full compressor shown in the compressor map in Figure 3.9. About half of the total pressure ratio of 4.2 is generated by the mixed-flow section, though with moderate efficiencies only. The axial stages raise the efficiency of the full compressor to a respectable 82%. The flat speed line characteristics permitted

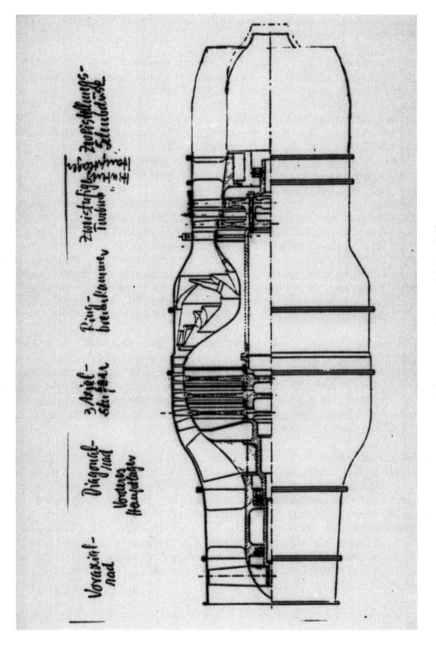

Fig. 3.8. Layout of the He S 011 (1942).

use of a two-position jet nozzle and a simplified engine control without variable geometry or compressor bleed. Figure 3.10 shows the compressor installed in the engine.

Compressor testing became fairly routine except for an occasional incident. To confirm the mechanical integrity of compressor and instrumentation, our procedure called for a repeat of a test point before closing the test series. In one case, pressure and temperature did not jibe. Since vibratory and oil temperature levels were normal, we scratched our heads for an explanation of the discrepancy. A technician came to our rescue saying, "Doc, maybe that's it" and pointing to a handful of broken rotor blades he had found near the air exit outside the test building. He was right: We had completely wiped out one aluminum stage. That left the rotor fully balanced and oil temperature and vibration unaffected.

A nagging compressor problem was the correlation of the measured data with physically the same instrumented compressor when it was operated on the compressor test stand and then installed and operated in the jet engine. In the former case, the inlet to the compressor was throttled and the exit at atmospheric pressure. In the engine, with the inlet at ambient atmosphere, the compressor map shifted to the right, the massflow being up to 9 percent larger than with the throttled inlet. Measured compressor efficiencies in the engine were lower than those with throttled inlet. Testing the compressor in Dresden under various inlet conditions up to the full pressure levels confirmed the Zuffenhausen data.

BMW, Junkers and DVL observed similar discrepancies between inlet and exit throttling of compressors, though the data was hardly consistent. Reasons for these deviations may be the difference in Reynolds number and effective air turbulence at the compressor entrance. Though not fully understood, one factor was fairly clear: Saving mechanical drive power by applying extreme inlet throttling and/or exit suction will lead to "measured" compressor characteristics which bear scant resemblance to those experienced under engine conditions.

As mentioned above, the mixed-flow wheels could stand further improvements. Our Zuffenhausen stand having been damaged in an air raid, we tested modified wheels in March 1945 in Göttingen. There I vividly observed the dichotomy between industry and academic research institutions. We were daily threatened with enemy interference from the air, suffered from air raids on Stuttgart, and were still frightened by the "Black Thursday" October 1943 attack on Schweinfurt (which my brother, his wife and two teenagers fortunately survived) and the destruction of Dresden (where we had wrapped up our compressor testing a few weeks before). In contrast, the Göttingen people were isolated from the rough and tumble world in a cloistered, tranquil oasis.

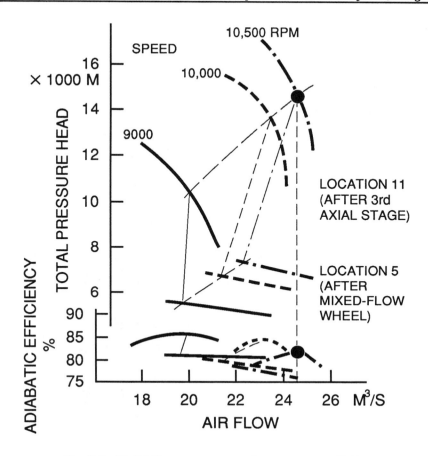

Fig. 3.9. He S 011 compressor performance map (1943).

I can illustrate this contrast with one episode. Late one morning I received an urgent call to come to Encke's office. Rather disturbed and agitated, he confronted me with the following: In the men's room he had found Herr Fischinger (one of my test engineers) standing barebreasted in front of a mirror and shaving his beard. I calmly replied, "So what. He was probably working late last night and wanted to face the world clean shaven. Come on, let's forget the incident and go to lunch." To which he said, "We can't yet, we have to wait for the buzzer to sound!" (This anecdote has a sad ending. On his return from Göttingen to Stuttgart, Herr Fischinger was killed in a minor unannounced air raid. At his gravesite memorial service, I remembered the shaving incident with tears in my eyes. C'est la guerre! C'est la vie!)

Fig. 3.10. Compressor installed in He S 011 engine (1943).

Combustor

The inheritance of combustor development was my worst assignment. It started with my concern over too much sheet metal in the primary and mixing zones, and over experiments with small sectors of the annular combustor. In my opinion, the sector side walls affected the whole process to such an extent that the experimental results were close to useless. After engine runs had confirmed this opinion, everybody — up to top management — started to suggest an appropriate combustion scheme which we nicknamed "Sunday chamber," "Versuchskarnickel" (Guinea pig), "Ratzenschwanz" (rat's tail), and worse.

The time was apparently not ripe for an annular combustion system of short length, low pressure drop and high burning efficiency, good reliability and adequate life. To the end, the 011 combustor remained an oddity, with its shovels to force mixing of air and fuel and to generate a uniform peripheral and a desirable radial temperature profile. The combustor barely fulfilled its purpose but was far from a page of glory.

Turbine

For efficiency reasons, the turbine was laid out with two axial stages, originally with solid blades, only the first nozzle vanes being air-cooled. Having no hot turbine test rig, the turbine was run in engine tests starting in September 1943, twelve months after go-ahead.

This configuration, a convenient expediency at that time due to the lack of air-cooling technology, experienced two types of failures: stress-rupture of the first and fatigue of the second rotor blades. We traced the latter to the four struts of the rear bearing support and eliminated them by spacing the struts unequally at angles, thus minimizing the forced excitations which were in resonance with the second stage rotor blades.

The rupture failures were, of course, the result of trying too hard to achieve the design thrust at a time when compressor and combustor were still far from their specifications. Relying on my experience with air-cooled turbocharger blades, illustrated as the *Faltschaufel* of Fig. 2.5, I proposed and scaled its derivative, the *Topfschaufel*, to the He S 011 size, see Fig. 3.11 (variously called tubular or bootstrap blade). Its major advantages were ease of manufacture without jeopardizing mechanical integrity and aerodynamic performance, and minimum need for strategic materials.

The blade consisted of an integral airfoil and root which was held in the disk with an axial pin. This attachment minimized the blade stresses, gas bending and

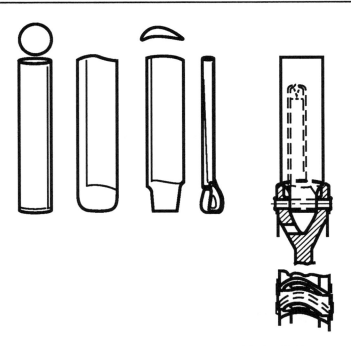

Fig. 3.11. He S 011 hollow turbine blade (1943).

centrifugal forces acting independently. A sheet-metal insert served two purposes: It ensured the proper distribution of the cooling air to the leading and trailing edges, thereby minimizing the amount of cooling air, and it damped blade vibrations. A sheet-metal cover between adjacent blades was provided with a small air gap to accommodate the change in angular blade positions with speed and pressure level. To eliminate excessive air leakage, each blade was equipped with a small collar. The covers carrying the collars were attached to the pins on the outside of the turbine disk. Manufacture started with a circular plate of austenitic chromium-molybdenum sheet steel, from which a closed-end tube was drawn in several stages with intermediate heat treatments. Its wall thickness diminished from 2 mm (0.079 inch) at the root to 0.45 mm (0.017 inch) at the blade tip in a manner to match the blade stresses with the prevailing radial temperature distribution. The tube was coined to the airfoil and root shape and then finish-machined. Both turbine stages were equipped with this rotor blade type and also with air-cooled stator vanes.[10]

The air-cooled turbine, shown as a cutaway in Figure 3.12, proved essentially free of mechanical, durability and operational difficulties; it readily allowed a rise in turbine inlet temperature and thus growth in engine thrust. Its efficiency of 85 percent was remarkable, too. The turbine presented the masterpiece of the He S 011, of which I could be justly proud. It contributed even more than the unorthodox compressor to the chief interest and attraction of the Allied engineers to this second-generation jet engine.

Performance of gas turbines is increased by, among other factors, the maximum cycle temperature. This requires heat-resistant materials or cooling of the hot components. Except for cooling of the combustion chamber walls and of the first turbine nozzle vanes, air-cooling is associated with losses in performance and overall efficiency. These detrimental effects increase with decreasing mass flow; they can be eliminated by the use of ceramics for the gas turbine hot section.

In Germany, this possibility was recognized from the beginning of turbocharger and jet engine development, and R&D efforts were conducted mainly on behalf of government agencies and institutions. Our Hirth Motoren management was reluctant to divert resources to these efforts, partly because it hesitated to add to the difficulties already caused by the spark-plug ceramic. Furthermore, the hot turbine inlet scroll of the DVL/Hirth turbocharger was originally insulated from the bearing housing with a ceramic separator which frequently cracked, thus contributing to the animosity against ceramics in engines.

Ceramic components for gas turbines offer a variety of advantages. Made from abundant domestic "dirt" and, therefore, supposedly dirt-cheap, they possess low specific weights resulting in low centrifugal stresses and low overall weight, and are allegedly resistant to corrosive gases. Most important, they permit high cycle temperatures without the parasitic losses associated with air-cooling.

Disadvantages of ceramic components were their great brittleness and low material damping coefficients, both presenting major obstacles. Combinations of ceramics with metals were investigated to overcome these shortcomings.

Two rotor blade designs with monolithic ceramics were pursued: an orthodox one with a dovetail attachment to the disk and one providing more acceptable

---

[10] WMF, the world-famous company for cutlery and silverware, manufactured the tubes. Their technologies were very advanced, but only in terms of processing materials. Appendix 3.1 explains how we familiarized them with the rigorous treatment their products experience in jet engines, in contrast to their tableware at formal dinners.

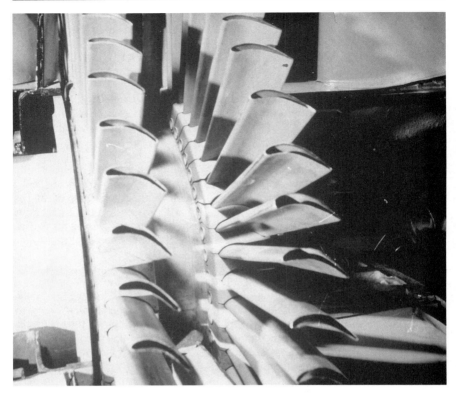

Fig. 3.12. Cutaway view of air-cooled He S 011 turbine (1944).

compressive stresses consisting of an outer "cold" metal ring carrying the cantilevered blades. Specimens of each, together with a variety of metallic counterparts, fill my memorabilia boxes.

The Lilienthal Society arranged a two-day conference on ceramic gas turbine components at the Carl Zeiss plant in Jena in November 1944. I vividly remember the heated discussions of the proponents and opponents leading to fruitless arguments but to no consensus.[11]

---

[11] Forty years later, similar conferences held in Dearborn, MI, Watertown, MA, Bad Neuenahr, West Germany, and Osaka, Japan, revealed almost identical dichotomies. Ceramic gas turbines apparently still have a long way to go, and they won't be inexpensive either. Recently, structural ceramics established a small foothold with turbochargers for automobile engines.

An air-raid alarm interrupted the Jena conference. As the factory shelter was short of space, some of us left the plant and enjoyed the mild Indian summer sun, lying on our backs at the edge of a forest. Hundreds and hundreds of bombers converged in waves toward their Leipzig target 50 miles away. Some B-17s flew so low we could see with the naked eye their four Moss turbochargers and their bomb doors — fortunately still closed. In the distance, we observed half-a-dozen fireballs of burning bombers but were unable to discern whether these were caused by flak, conventional, jet or rocket fighters.

After a few more briefings at the Air Ministry, I closed my files on ceramics. Lo and behold, these were one of the first ones I had to divulge to the British engineers who interrogated me shortly after VE day.

Jet Engine He S 011

Development and testing of the full He S 011 engine proceeded in stages, depending on the availability of improved compressors and the hollow-blade turbine. In December 1944, 27 months after contract receipt, best performance figures were achieved as follows:

Thrust 1333 kp (2940 lb, 13.1 kN) at a speed of 10,205 rpm, pressure ratio 4.2, mass flow 29 kg/sec (64 pps), turbine inlet temperature 775°C (1427°F), fuel consumption 1.35 kg/kp-hr (1.35 lb/lb-hr).

A photograph of the He S 011 engine, while on exhibit at the Air Force Museum in Dayton, is shown in Figure 3.13; this cutaway engine is on permanent display in the Crawford Auto-Aviation Museum in Cleveland, Ohio.

The performance milestone triggered the release of the production drawings, the first mass-produced engine being scheduled for May 1945.

On April 9, advance of the Allied forces toward Stuttgart caused evacuation of the Heinkel-Hirth team to BMW, Munich, and at the end of April, to Kolbermoor in southern Bavaria.

Propulsion Systems for Long-Range Aircraft

Simultaneous to the development of jet engines, alternative concepts were evaluated. Simple turbojet engines (TL = Turbinen-Luftstrahltriebwerk) exhibit proverbially high fuel consumptions, both theoretically and practically. At the beginning of their development, this drawback appeared to make them suitable only for fighter aircraft. Proposed solutions to increase the overall propulsion efficiency and aircraft range consisted of ZTL (Zweikreis-TL (second stream) = turbofan), PTL (Propeller-TL = turboprop) and ML (Motorrückstößer-Luftstrahltriebwerk = engine jet propulsor). All three types were pursued in Germany.

Fig. 3.13. He S 011 engine on exhibit at the Air Force Museum in Dayton, OH (1978).

ZTL DB 007

In 1939, Daimler-Benz received a government contract for a turbofan. Professor Fritz Leist (previously with DVL) was in charge of the project. His proposal, designated DB 007, presented a fairly complicated design, consisting of two compressors with counterrotating drums à la Weinrich, the main compressor with eight stages for a pressure ratio of 8, its outer drum carrying three stages for the second airstream and the cooling air for the partial-admission turbine. The single-stage axial turbine drove both drums via a planetary gearbox. Afterburning in the bypass and an adjustable jet nozzle completed the arrangement.

Actual development started in 1941. The first experimental engine was run in 1943 and finally achieved full speed. The long development time and the uncertainty about the viability of this complex aerothermodynamic and mechanical arrangement induced the Air Ministry to abandon the project.

PTL DB/He S 021

In view of the development progress of the Heinkel-Hirth He S 011 engine and DB's free capacity, the Air Ministry ordered the development of a turboprop, designation DB/He S 021 based on the 011 core engine. DB carried out various designs but no experiments.

ML He S 50

An ML was considered as a replacement of the recip/propeller system, posing fewer development risks than a turbojet. Figure 3.14 shows the cross-section of one of the He S 50 versions.

It comprised these major components: a reciprocating two-stroke engine driving the main (axial) compressor for the cooling air and an additional (radial) stage for the engine charge and scavenge air; a mixing chamber which included a fuel injector for thrust augmentation; and the jet nozzle. Heat input to the propulsion air was from the compressor, from the cooling air and the exhaust of the reciprocator, and from the auxiliary burner for afterburning.

The reciprocator was based on studies and experiments with small cylinders which Professor Dr. Wunibald Kamm (1893-1966) had conducted at the TH Stuttgart.[12] They offered advancements in air-cooling and high power concentration without detonation. One 48-cylinder engine was actually built and run; it had been nicknamed *Stachelschwein* (porcupine).

---

[12] Kamm was world-famous for his pioneering work on passenger cars, particularly steering, stability, and inner and outer aerodynamics. His full-size five-seat K-Wagen achieved the extremely low drag coefficient of 0.23. It was a familiar sight on the streets of Stuttgart and on the autobahn.

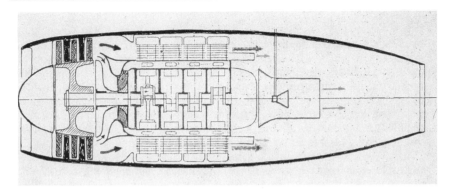

Fig. 3.14. Cross-section of the He S 50 (1939).

After Heinkel's transfer of the propulsion groups to Zuffenhausen and Müller's departure, I was put in charge of this project. Impressive were the calculated specific fuel consumption figures which were, compared to the He S 011, at sea level some 50 percent, and at 10 km altitude some 35 percent lower.

Experiments with two cylinders of the reciprocator revealed difficulties caused by the combination of high pressure and temperature. The pistons particularly could hardly stand up to this severe environment. Cooling of the piston by blowing air through a tunnel between dome and piston proper proved insufficient. This and other problems as well as the progress made with turbojets led to abandonment of the project.

In retrospect, the He S 50 powerplant was a more viable jet propulsor than the Italian Caproni-Campini engine/two-stage radial compressor configuration which had made headlines with its Milan-Rome flight in 1941.

It is conceivable that the ML concept will be revived for General Aviation aircraft in the not too distant future.

Early Jet Engine Development - Postscript

In contrast to the aircraft reciprocating/supercharger/turbocharger field where our technical information from abroad was continually updated with captured material, no such convenient windfall existed for jet engines. Our intelligence gathering was restricted to publications of our foreign competitors that we obtained via neutral countries, and to international professional magazines, e.g., *Interavia* of Switzerland. As a bonus, we had once acquired a photostatic copy of an early edition of G. Geoffrey Smith's (1885-1951) booklet "Gas Turbines and Jet Propulsion for Aircraft," which described the pioneering work of Frank Whittle (1907-). That was all; in the parlance of the late 1960s, we essentially "did our own thing."

After the end of World War II, two versions of the early jet engine development emerged: one by the professional press and societies which was based on captured engine components and engines[13] and their drawings, and on our interrogations and reports; and the other one destined for the public-at-large and the world, where our contributions and progress were downplayed or simply ignored.

King George VI knighted Frank Whittle in 1948, and the British Government honored him with a tax-free stipend of 100,000 pound sterling, a considerable amount of money compared to the annual salary of 5000 pound of the Prime Minister and to my own of 1000 pounds.

During the 1951 Festival of Britain exhibition in London, the official booklet "The Story of the British Gas Turbine" contained the following paragraphs:

> Outside fairly recondite engineering circles, the gas turbine was scarcely heard of before the Second World War, and a great deal of work done during the war was necessarily secret until it was possible to announce to the world the success of Frank Whittle's experiments resulting in the operation of the first successful jet aircraft.

> The Germans, although they flew (in an experiment which was abandoned) the world's first jet aircraft, were hampered seriously in their development by inferior turbine blade materials which led to inefficient results.

In 1966, Sir Frank Whittle and Dr. Hans von Ohain met for the first time in the U.S. Whittle was then consultant to British jet engine manufacturers and von Ohain was the Chief Scientist of the Air Force Aerospace Research Laboratories. They informally discussed their views and experiences. Further details became known at an "official encounter" of these two specialists on invitation of Philippe O. Bouchard, Brigadier General, USAF, Vice Commander Aeronautical Systems Division. These give-and-take conferences climaxed on May 4, 1978, with a

---

[13] The Allies' first lucky scoop happened in February 1945, when an Arado 234 made an emergency landing in the Rhine valley with two operable Jumo 004 engines.

session at the Air Force Museum including public interviews with Sir Frank and Hans, whose former co-workers were privileged to be introduced, e.g., Helmut Schelp and me (see Figure 3.15).

In frank presentations and free discussions, the history of the early jet engine development became clear. Hans's great fortune in obtaining Ernst Heinkel's enthusiasm and support of his jet propulsion ideas and Sir Frank's misfortune of failing to gain such a helpful proponent early enough seemed particularly significant. For these reasons, Whittle was denied the fame of being first with a jet-propelled flight. His patent application was three years *ahead* of von Ohain's *idea*, but his first *flight* some twenty months *behind*.

Toward the end of the lively Q&A period an attendee asked, "For a considerable time, I have listened to both of you (Sir Frank and Hans), digested all you said but I'm still puzzled who was actually first."[14]

The answers given to the jet engine "firsts" are presented in Appendix 3.2, including the milestones of the American development, with the splendid recovery after the 1938 misjudgment.

The parallel independent jet engine development in Great Britain, the U.S. and Germany is full of similarities and also basic differences. In 1939, the producible heat-resistant alloys were about equal on all sides. Then the availability of important alloy ingredients rapidly decreased in Germany; nickel, cobalt, manganese, copper, silver, etc., had to be substituted with chromium, molybdenum and columbium (called Niob). With this *Ersatz*, the material properties decreased at high temperatures and air-cooling became a necessity, both for turbochargers and jet engines. For a rise in gas temperature and thus thrust and propulsion efficiency, the British would simply increase the nickel content of alloys, whereas we had a more difficult task, namely to use inferior materials, cool them with air and use the

---

[14] This question and its answers confirmed my axiom: An invention or event can often be made a "first" by supplementing it with an adjective or other qualification. An example is the helicopter story. Encyclopedias record the actual inventions and events; almanacs and other publications and the general public in general assume that Igor I. Sikorsky (1882-1972) invented the helicopter. Even the reputable New York Times expressed this opinion in 1975 with the statement: "The vertical lift machine, developed in concept by Leonardo da Vinci in the 15th Century, was first raised inches off the ground in 1939 by Igor I. Sikorsky." I reminded the editor of Hanna Reitsch's flight with the Focke helicopter FW-61 inside the Deutschlandhalle, which had preceded by over one and a half years Sikorsky's first flight with his VS-300. The FW-61 had reached an altitude of 2349 meters (equal to 96,023 inches to be exact). The replica of Sikorsky's first helicopter in Bridgeport's Museum of Arts and Industry was then marked "The first *practical* helicopter." (Focke's helicopter with its twin-rotor was inherently stable; Sikorsky finally achieved stability in free flight with a single main rotor. In 1971, Hanna Reitsch and Igor Sikorsky enjoyed a most amicable encounter discussing the early days of aviation.)

Fig. 3.15. Encounter at the Air Force Museum. (From left) Helmut Schelp, Sir Frank Whittle, Hans von Ohain, and Max Bentele (1978).

cooling air with high effectiveness. Our materials had, however, the advantage of a higher elongation, which helped in reducing stress peaks and in avoiding catastrophic turbine failures during development testing. Rubbing of the blade tips would produce visible sparks and an alert technician was able to save the turbine from destruction by easing the fuel throttle.[15] Air-cooling finally eliminated the radial-inflow turbine despite its higher efficiency potential.

In the quest for high thrust per frontal area, Germany favored axial-flow turbomachinery. Axial compressors used in industrial plants were, however, too cumbersome and heavy.[16] New aerodynamic investigations were proposed and conducted. Anxious to obtain useful results on time, von Ohain, as did Whittle,

---

[15] The detrimental effects of low-elongation materials are even more severe with ceramics. Engineers jumping into their applications had to learn that fact the hard way.

[16] See, in this connection, the same concern of the Committee of the (U.S.) National Academy of Science.

selected radial compressors, and both built and flew successfully with these. As late as 1950, at a meeting of the Royal Aeronautical Society, the merits of radial and axial compressors were still hotly debated and defended by representatives of De Havilland and Rolls-Royce, respectively.

Combustion of jet fuel was of major concern to Whittle. His W 1 and subsequent engines dramatically illustrate this point (see Figure 3.16). The reverse-flow combustor cans increase the overall engine diameter over that of the impeller diameter by a whopping 80 percent.

In Great Britain, the axial turbine with solid blades was more amenable to optimizing the airfoils and radial twists than our air-cooled hollow blades. In a single stage, the British design converted high pressure ratios to mechanical work with remarkably high efficiencies.

In Germany, the use of inferior materials and the need to tailor the components to mass-production with mostly unskilled labor had detrimental effects on reliability and life expectancy. The Air Ministry deemed that acceptable in view of the hazards and casualty rates of the wartime operation of the engines.

The 40th anniversary of the first jet flight was celebrated with a session at the National Air and Space Museum in Washington on October 26, 1979. Whittle, von Ohain, Franz and others presented papers which the Smithsonian Institution published in the book, The Jet Age - Forty Years of Jet Aviation.

One of my hobbies is amateur drawing and painting, particularly of greeting cards for birthdays and anniversaries. The recipients invariably appreciate the personal touch fitting the occasion. The celebration in the Smithsonian of the 40th anniversary of the first jet flight induced me to depict this event, which was exhilarating to Hans von Ohain and disappointing to Frank Whittle. The von Ohains enjoyed this greeting card sent for Hans' birthday in that year (Figure 3.17).

By 1987, the true story of the jet engine pioneering work performed in the UK, U.S. and Germany was well known worldwide to the professional and lay public alike. An article in the *Reader's Digest* carried the headline, "They Created the Jet Age," referring to Sir Frank Whittle and Dr. Hans von Ohain, all working in this novel and fascinating field in Great Britain, Germany and the U.S. There were enough national and international honors, laurels and awards to go around, and they are still coming.

The 50th anniversary of the first jet-powered flight was celebrated in the United States and in West Germany. The American Institute of Aeronautics and Astronautics (AIAA) sponsored a one-day symposium on August 23, 1989, in Dayton, Ohio,

Fig. 3.16. Whittle's W 1 jet engine (1941).

and the Deutsche Gesellschaft für Luft und Raumfahrt (DGLR, German Society for Aeronautics and Astronautics) organized a two-day International Symposium "50 Jahre Turbostrahlflug - 50 Years of Jet-Powered Flight" held on October 26 & 27, 1989, at the European Patent Office and the Deutsches Museum in Munich.

Magda and I were honored guests of the DGLR Society and I was privileged to present a paper at the symposium on "Das Heinkel-Hirth Turbostrahltriebwerk He S 011 — Vorgänger, characteristische Merkmale, Erkenntnisse (The Heinkel-Hirth He S 011 Turbojet Engine — Its Predecessors, Characteristics and the Lessons Learned from It)."

The speakers immediately preceding my talk had presented a very scientific and rose-colored picture. The SRO audience encompassed the whole spectrum from the early pioneers and pilots, government officials, academics, industry leaders and

Fig. 3.17. Birthday card for Hans von Ohain commemorating the 40th anniversary of the first jet flight.

employees to engineering students, relatives and friends. I accordingly adjusted the preprinted text by adding the human endeavor in the real world as follows:

> "To present unpublished design features and test results of the then most powerful jet engine is for me a high honor. I also appreciate it because my first excursion from the nearby Ulm High School in 1926 had focused on Munich and its Deutsches Museum. Describing briefly the engine's predecessors, I concentrated on the design and development of the He S 011. The targets for this first engine of the Power Class II were increases of the thrust by 30 to 50 percent, of the pressure ratio by 35 to 90 percent, and of the power/weight ratio by 20 to 40 percent. There was another clause: Start of production 30 months after receipt of contract. That was a "tall order," as one would say in the States. The *Diagonalrad* (mixed-flow wheel) was adequate, but its improvement had to be postponed since our best computer system only consisted of 20-inch slide rules and tables for logarithms and other functions.[17]

> "After twelve months a jet engine was assembled with the then available components. The test results were, to put it mildly, disappointing and sobering. The second turbine stage suffered from vibratory failures, and the first stage already reached its thermal limit at half of the design thrust. The jet stream contained sparks which indicated incomplete combustion and metal scales. An attempt to exceed 1000 kilopond (2200 lb) of thrust, ended in thermal breakage of a first-stage blade.

> "With successively improved components the target performance was reached and exceeded in December 1944.

> "After summarizing the lessons learned from this development (as described in the relevant book sections above), I paid tribute to the noteworthy He S 011 pioneer work, which was the result of the engineering expertise, excellent cooperation and dedication of the Heinkel and Hirth teams. Some members participate in this symposium; I gratefully acknowledge their efforts to be here, and their devotion."

The moderator thanked me for my forthright talk and for remembering the working conditions and team spirit of that time.

---

[17] This apology was greeted with laughter and applause. Several subsequent speakers took pleasure repeating it.

For the convenience of the Bonn establishment, a reception, mini-symposium and dinner, sponsored by the DGLR and the Club der Luftfahrt von Deutschland, were held afterwards at the Godesburg, the landmark of Bad Godesburg. As did other honored guests, Magda and I stayed at the Rheinhotel Dreesen, famous as the place of the first encounter of the British Prime Minister Neville Chamberlain (1869-1940) and Adolf Hitler (1889-1945) in the Fall of 1938. This meeting had led to the Munich Pact for which Chamberlain had coined the phrase "Peace for our time." We reminisced about our honeymoon automobile trip along the Rhine river which had preceded these historic events, and the turbulent times following them. Now we again admired and enjoyed the scenery of the wide river, its pleasure and cargo boats, and its idyllic banks.

Appendix 3.1

Turbine Rotor Blade Environment

The turbine rotor blade of a jet engine represents the highest stressed part in the hot section, and the most critical. Fracture or loss of a blade will cause a loss of power at best and an engine shutdown or loss at worst.

The design engineer describes the blade operating conditions with complex diagrams, coefficients and numbers, and determines their relationships to the material properties. Contributing and stress-inducing factors are the material temperature along the turbine radius and across the blade airfoil section at each radius; the mechanical forces caused by the turbine rotation, the gas stream and the radial stacking of the blade sections; nonuniform temperatures of the individual cross-sections causing thermal stresses; and finally, the varying engine operating schedules, starting and shutdowns, leading to the low-cycle fatigue damage and failure of the material.

To explain these diagrams and their critical items to an engineer familiar solely with the design and production of high-class cutlery would have been a tedious and unrewarding task. So we chose a simpler and more effective indoctrination. We showed these engineers an engine mounted on the test stand. They could see and touch the blades held with a pin in the turbine disk. Then we provided them with earplugs, led them to a location out of the direct jet stream and started the engine. With increasing engine speeds the rotor blades became indistinguishable. The turbine guide vanes started to glow, and the jet exhaust roared to an almost unbearable noise, blowing dust onto their goggles and one hat off. The visitors were impressed. Deceleration and snap acceleration increased their excitement about experiencing a jet engine in operation. Though they were unable to fathom the significance of the stress and temperature diagrams, they would clearly get the message: The turbine blades they produce had better be excellent, otherwise the engine will fail, and with that their reputation will irreparably suffer.

Appendix 3.2

Jet Engine Firsts

1921	French Patent granted to M. Guillaume. Axial compressor, axial turbine. Not pursued.
1930	Frank Whittle applies for British patent. Axial-cum-radial compressor, in-line combustor, axial turbine.
1935	Hans von Ohain runs rudimentary model, back-to-back radial compressor/turbine (Fig. 3.1).
1937, April	von Ohain runs hydrogen-fueled turbojet (Fig. 3.2).
1937, April 12	Whittle runs liquid-fueled turbojet. Dual-entry radial compressor, single-can combustor, axial turbine.
1939, Aug. 27	Flight of jet-propelled aircraft Heinkel He 178 (Fig. 3.4), von Ohain engine He S 3 B (Fig. 3.3).
1941, April 2	Flight of twin-jet fighter Heinkel He 280, von Ohain engine He S 8 A (Fig. 3.5).
1941, May 15	Flight of Gloster E 28/39, Whittle engine W 1, dual-entry radial compressor, reverse-flow combustor cans, axial turbine (Fig. 3.16).
1941, Oct. 1	General Electric Company (U.S.) receives disassembled Whittle engine W 1 X and drawings of Whittle engine W 2 B.
1942, April	GE bench-tests its I-14 engine based on W 2 B engine.
1942, July 12	Flight of twin-jet fighter Messerschmitt Me 262, Junkers Jumo 004 A engine.
1942, Oct. 2	Flight of Bell XP-59 A, GE engine I-14.

One has to take the people as they are;
There are no other ones.
Konrad Adenauer (1876-1967)

Chapter 4

End of World War II

Interregnum

In early 1945 the approaching battlefront and frequent day and night air raids gave us Heinkel-Hirth engineers plenty of time to meditate and discuss our professional future. The war's end would essentially leave us three options: (1) quit industry altogether and try a more mundane service job; (2) stay in industry but concentrate on conventional consumer goods and agricultural equipment; or (3) continue with gas turbines with the hope and wish that our knowledge and experience would be appreciated somewhere.

The choices of my colleagues were about equally distributed among these categories. I myself selected the third option, out of professional pride, self-esteem, opportunism or however people may see it.

In early April 1945, the Heinkel-Hirth major departments were ordered to abandon the Stuttgart premises and travel by train to Munich, Bavaria. Hubert and I chose to think of our family first and jet engines second, and proceeded by bicycle to our loved ones at Brachbach. A few miles away from the village, a German motorcycle patrol stopped us with the surprising question, "Do you know where the Americans are?" To our negative answer, they said, "We think they encircled us." After this strange encounter, we were happy to reach Brachbach and our families. Apparently in no-man's land, we enjoyed a good night's rest.

The next day, Sherman tanks approached Brachbach from the west and, together with infantry men, armored and ambulance vehicles, occupied it. My presence caused some suspicion but I explained it to their satisfaction. The quietness of the evening was suddenly disturbed by the rattle of their radios. All GIs jumped on their vehicles and left eastward. We hugged each other; for us this terrible war was over! After supper, my family and I retired to our room, the former dance hall on the second floor.

During the night, our youngest daughter became restless and started to cry. My wife comforted her back to sleep, then heard a staccato noise from the street and took a look. A platoon of SS soldiers was marching by, eastward. Our joy of the evening was shattered; we were in the midst of the war again! For a couple of days, we had no idea about our situation. Then Wehrmacht contingents bivouacked around the hamlet, together with their equipment, vehicles and guns. One evening, a soldier approached me with "Hi, Doc, remember me?" He was one of my technicians before he was drafted in 1944. His story was: "I am with a Nebelwerfer (rocket launcher, nicknamed Stalin-Orgel), our route brought us from the English Channel via Belgium and the Rhineland to here. They (the enemy) hate us more than anybody else. As long as we don't fire, Doc, you, your family and neighbors, are pretty safe. If we do fire, the Lord be with you all!" Well, one night an earsplitting noise awoke us; they did launch quite a few rounds, then left with all other units.

So we were on our own again! Shortly before noon, an American observation plane circled Brachbach. Since no military equipment and personnel were there anymore, I felt safe and attended to some chores in the yard. When I heard distant noises of fighter planes, I recalled my technician's warning and shouted aloud, "Take cover!" and we all ran to the shelter, a huge stone-arched cellar, deep underground. We had hardly reached there, when the house and barn were hit with gunfire and incendiaries, the neighborhood also with bombs. After it became quiet, we emerged from the cellar to find the house and barn burning. We were able to save the house but not the barn. We counted some 150 shot holes in our house alone, ten destroyed buildings and a dozen bomb craters in the hamlet. We also counted our blessings, though: no major human injury nor casualty, just a few wounded and dead cows and fowl.

For the night, we took no chances. We and our neighbors, two dozen men, women and children, spent a peaceful time in the cellar. Shortly before daybreak, the cellar door opened quietly. Two GIs, standing there, wondered who we were and whether we were hiding German soldiers. I assured them of the contrary and asked about their wishes. They simply answered, "A few fresh eggs, dark bread, and some jam!" We obliged cheerfully and were happy that the war hostilities were now really over for us.

With this interlude, I never made it to the village of Kolbermoor outside Munich, where the Heinkel-Hirth team had ended up and encountered the Americans a few weeks later.

Our "ideal" country retreat had suffered from the imponderables of war. The past week had made it abundantly clear to me that there was little an individual could do to circumvent these. Still, luck was with us and we were able to prepare for our new life.

All usual communication with the world had suddenly ceased: no electricity, and thus no radio, no telephone, mail, newspaper, public transport, bus, train, no nothing. We lived on an isolated island, although we had plenty of work and food. Due to the lack of transportation, people in the cities were starving, whereas we had to bury our surplus meat, eggs and dairy products in the ground to reduce the danger of infection and disease.

Slowly our sources of information opened up through word of mouth, including rumors, chats with GIs, reading their *Stars & Stripes*, and so on. We thus learned about President Roosevelt's death, Hitler's suicide, Germany's surrender and VE-Day on May 8.

First things first, I tried to obtain personal contact with our blood and other relatives and friends. I was fairly well equipped for this task: a working knowledge of the English language, a sturdy bicycle, a road map of southern Germany in my head, and a company I.D. card with my photo and 1945 sticker.

Communication with my brother Carl was easily established. He had emigrated in 1925 to the States, and lived in Boise, Idaho. Correspondence had, of course, ceased after Pearl Harbor. A sergeant volunteered as a go-between and transmitted a letter to my brother. In his return military mail, he expressed joy and satisfaction that we were all right and that I was trying to find the other family members. For this purpose, I bicycled to Jungingen, 70 miles to the south. Such travels were officially verboten, the occupation authorities permitting only visits to neighboring villages. When MPs stopped me for identification and questioned me about my trip, I simply answered I was coming from *that* village and going to *that* one. This procedure always worked.

Avoiding cities and towns, I once saw a Sherman tank lying under a bridge in a brook. Pointing to the tank, I inquired of a local man about heavy fighting in this neighborhood. He replied, "There was no fighting at all. The tank commander and driver had ignored at the bridge the red-circled sign '5 t,' evidently unfamiliar with European standards. The tank was obviously far heavier than the inscribed 5-ton limit, and the bridge collapsed. I felt sorry for the guys."

Ulm, badly damaged by air raids, had hoisted white flags of surrender, as had Jungingen. Since my last visit there in February, some air-raid casualties and damage were suffered but my loved ones were okay. My sister Gretel and her family, evacuated from Ulm, were still in grief about her only son who was killed in action in Hungary. Similarly, I found my wife's family with relatives outside Heilbronn; they had survived a heavy air raid and lost their house, but were in fair condition. My brother and his family in heavily damaged Schweinfurt, the ball-bearing city, were also all right.

On a further investigative journey, I came to Bissingen/Enz, 14 miles north of Stuttgart, then under French occupation. The Maschinenfabrik G.F. Grotz, a local factory belonging to the Heinkel group, was only slightly damaged and already in haphazard production of household goods, such as a gadget for making homemade noodles, cutting tobacco leaves for cigarettes, and other thingamajigs. At that time, Grotz was without top management. Its co-owner had not yet returned from the eastern front and Ernst Heinkel himself was retained for questioning at Wimbledon, England, together with scores of other heads of military/industrial firms. Assuming his approval *in absentia*, I became General Manager and Chief Engineer of Grotz, ready to receive the visit of French officers. They brought a cardboard box of spare parts for their American-built Jeeps which they asked me to manufacture at Grotz or procure from other companies. With a straight face (and a smile inside), I expressed surprise about this obviously uneconomical way to keep their Jeeps going. Wouldn't the Americans be glad to dispose of their surplus of spares by *giving* them to their allies? The officers mentioned ongoing disputes about the territories to be permanently occupied by the French or by Americans, among other reasons for their request. On second thought, it was hardly the time and place for me to argue the point. After all, it would keep the factory going, as well as bring me back to industry and my family from the Brachbach hinterland closer to Stuttgart, our familiar neighborhood. With a handshake I happily agreed to put their procurement program in place. I ordered laboratories to analyze the materials of the parts, made shop drawings of the low-tech items for Grotz (wheel nuts and bolts, valve and valve spring retainers, among others), and located companies who would supply the rest, namely ball and roller bearings, shaft seals, engine valves and valve springs. This lucrative business in Jeep spare parts continued to the satisfaction of Grotz and the French Air Force long after Bissingen was included in the American occupation zone and I had left for high-tech ventures.

My Hiatus in the Western Occupation Sectors

After my short stint at Grotz, I was asked in July to rejoin Heinkel-Hirth in the Stuttgart plant. I then learned the course of events which I had missed since my detour in April to Brachbach.

Toward the end of the European war, the Allied powers had concentrated their intelligence on German science and technology and on those who were part of it. This gathering of information and of prospective candidates for hire was carried out in several waves. The first consisted of Allied Government officials assigned to this task; the second included engineers from their defense industry. Much later, representatives from their commercial enterprises were permitted to talk to us, and finally people came from small Allied nations. Unallied and unfriendly nations were trying to pursue these activities underground and achieved a considerable amount of success.

Typical examples of the first wave were Ernest C. (Cliff) Simpson of the U.S Air Corps, S. T. Robinson of the U.S. Navy and Reginald G. Voysey of the British Whittle Group.[1]

Impressed with the advanced features and performance of our He S 011 jet engine, the Americans wanted to have these verified with hardware. On German military orders, all experimental engines had been made inoperable or destroyed. As a consequence, the Heinkel-Hirth personnel were transferred back from Kolbermoor to the Stuttgart plant, which was reopened under the control of the U.S. Navy. Main assignments included written detailed reports on the engine and its components, building a dozen He S 011 jet engines for performance evaluation at sea level and at altitude, and for their eventual shipment to the U.S. and the UK. Also, we had to cooperate during interviews, interrogations and plant tours.

Initial interrogations included discussions with my British and American counterparts, to name a few: R.G. Voysey, R. Skorski of the Royal Aircraft Establishment (R.A.E.) Farnborough, Wilton G. Lundquist, Earl V. Farrar and others of Curtiss-Wright's Wright Aeronautical Division.

The discussions, often extending over several days, were very congenial and occasionally full of humor. We had the feeling that our interrogators were well-briefed on which subjects to dwell and which ones to avoid. Axial-flow compressors were a favorite topic. In their opinion, aerodynamic layout of axial blading was an art at best and alchemy at worst.[2]

Of the many areas of common interest, I particularly tried to impress our guests with my work on air-cooling turbines and on eliminating blade fatigue failures. The discourse on vibrations was very satisfactory to both sides, but my cooling developments did not fare so well. Our opponents' argument was simple: Our lack of high-temperature high-strength alloys forced us to this complication while their superior turbine materials permitted high turbine inlet temperatures without it. I countered with the suggestion that cooling their alloys would permit still-higher cycle temperatures and result in vastly improved performance. At that time and under these lopsided circumstances, our proposal did not sink in; it took almost ten years before air-cooling of gas turbines was seriously considered in the U.S. and in Great Britain.

---

[1] Simpson (1917-1985) interrogated von Ohain at Kolbermoor. He roamed the length and breadth of Germany, including the eastern parts which later came under Soviet Russian occupation. His illustrious career within the Air Force culminated in his position as Director, Turbine Engine Division, Air Force Aero Propulsion Laboratories, Wright-Patterson AFB. In 1980, he retired from Federal service. Voysey became manager of the C.A. Parsons Gas Turbine Department and was then elected to a high position in the Ministry of Fuel and Power.

[2] Shortly after the war, a book on axial compressor design published in Great Britain was immediately nicknamed The Theory of the Thousand Tests.

We were happy about these encounters. From their seeds finally sprouted the West German participation in NATO and AGARD.

Having essentially completed my reports and realizing the short-term duration of our work at Heinkel-Hirth, I tried to keep abreast of other developments in the gas turbine field.

The American camp for German engineers and scientists was located in Landshut in southern Bavaria. The standard contract offer to engineers included transfer to Wright Field, Ohio, with free board and lodging and $6 per day pocket money. After a get-acquainted period, the family would follow to Wright Field, free of charge. Quite a few would accept this deal, among others Helmut Schelp of the Air Ministry and Dr. Werner von der Nüll of the DVL (both later pursued small gas turbine and turbocharger development at the AIResearch Division of the Garrett Corporation), Dr. Anselm Franz and some of his associates at the Junkers company. (Franz later conceived and developed gas turbines for helicopters at AVCO Lycoming.)

We had no visits from French or Russian engineers. The USSR operated an office located near the Heinkel-Hirth plant, but that dealt mainly with prisoners of war and displaced persons. The French and Russians appeared to avoid contact with the American and British authorities, but practiced some cooperation between themselves.[3]

A blundered, tragic case was that of Siegfried Günter (1899-1969). In 1932, he and his brother Walter had conceived the aerodynamics and structure of the Heinkel He 70 in competition to Lockheed's Orion. The He 70 captured in a short time more than a dozen speed records and, with Lufthansa's name, "Heinkel Blitz," the admiration of the aviation community and traveling public worldwide.

Other novel Günter designs had followed, including the rocket-powered He 176 and the jet plane He 178. At war's end, Günter, an excellent engineer, exclusively dedicated to aircraft design and aerodynamics, worked for a short while with other Heinkel colleagues in a small office under American sponsorship in southern Bavaria. They wrote the history of Heinkel's jet planes and proposed new concepts for the future. Out of work in early 1946 he moved to Berlin, where the Soviet Russians got hold of him, recognized his unique capabilities and brought him to the Soviet Union.

The French, deprived of jet engine development during the war, established in their occupation zone on the eastern end of the Lake of Constance a meeting place

---

[3] A personal example: At war's end, the Russians transferred a brother-in-law of mine, who was their POW, over to the French. They had him work on a farm in southern France. I was fortunate to be able to send him from England a gift parcel with warm socks and underwear for Christmas 1947. He was finally released in March 1948, close to three years after VE day.

at Rickenbach, called Atélier Aéronautique de Rickenbach (ATAR). Dr. Hermann Oestrich of BMW, having declined the American standard contract offer, headed this establishment. He complemented his BMW team with engineers from the other German jet engine companies, Heinkel-Hirth and Junkers, thereby adding to his knowledge about his competitors' development results.

On a visit to Rickenbach, I discussed with Oestrich and his associates a possible position in the organization. I was offered a job with a fairly narrow range of responsibilities and eventually declined it.

Oestrich's design, based on an available preliminary BMW layout, satisfied the French authorities immensely. They transferred personnel and equipment to France. Their ATAR series of engines became world-famous and so did the company, Société Nationale d'Etude et de Construction de Moteurs d'Aviation (SNECMA).

Joseph Szydlowski (1896-1988), President of Société Turboméca located in southern France, also hired gas turbine experts from BMW, Daimler-Benz and Heinkel-Hirth, including some of my best engineers. Turboméca's initial project was an engine in the 15,000-pound thrust class for a transatlantic airliner, too large a step at the time. For this and other reasons, I did not pursue this lead any further. Turboméca later settled for engines in the low-power regime and became very successful with them.

An offer from Maybach Motorenbau, Friedrichshafen, involving turbocharger development work on a diesel engine for a French battle tank, was not to my liking either.

The last wave of interrogations included visits from a Dutch turbine firm and the famous American engine company Briggs & Stratton, Milwaukee, Wisconsin, among others. The latter was interested in our two-stroke engine work; they had only four-strokes in production. The Dutchman declined to make me a formal offer; he was afraid that the Americans would not release me, apparently unaware of President Roosevelt's Four Freedoms which included the freedom of movement.

The British, having developed jet engine technology neck-on-neck with us, established their headquarters near Braunschweig at the Luftfahrtforschungsanstalt Völkenrode (LFA, Aviation Research Establishment). They were keen to supplement their own knowledge with ours. At the Aerodynamische Versuchsanstalt (AVA) in Göttingen, all aspects of the German aviation R&D were preserved in monographs which finally amounted to 7000 pages.

Inquiring into the possibility of resuming testing of He S 011 compressors at the AVA (which was declined), I met there Dr. A.T. Bowden of C.A. Parsons, Newcastle-on-Tyne. He was interested in automotive gas turbines, and we discussed the work conducted at Heinkel-Hirth on this subject. Consequently, I was invited to study this terrestrial application further and submit a monograph.

The contents and noteworthy items of my monograph are delineated in Appendix 4.1.

My final proposal consisted of a simple-cycle regenerative gas turbine consisting of a gas generator, a free power turbine and a rotary regenerator. Ludolf Ritz of AVA had proposed this type of regeneration. It was derived from the Ljungstrom rotary regenerators that were employed in steel works for transferring exhaust heat energy from the furnace to its ingoing air. In the compressor field, I investigated the size effect which precluded, for low-power outputs, axial-flow compressors and brought Lysholm screw-type compressors into consideration. It gives me great pleasure and satisfaction that this automobile gas turbine concept was adopted worldwide in the late 1940s and is still being pursued in the 1990s. Recently screw-type compressors and expanders were again proposed for evaluation in small gas turbines.

My *Summary and Conclusions* read as follows:

"The basic aspects of the analytical and mechanical design of a vehicular gas turbine are discussed. Based on thermodynamic computations, promising development approaches are suggested and supplemented with numerous examples from practical experience.

"A skillful configuration will present torque characteristics more favorable than those of a reciprocating engine, a noteworthy operational advantage. A highly effective regenerator offers the potential to build a gas turbine powerplant with competitive fuel economies for the low-power outputs of road vehicles."

My AVA monograph contributed to my contract with the British Ministry of Supply (MoS) to assist in the development of a vehicular gas turbine.

During my work at the AVA, I had the pleasure and fortune to lunch occasionally with the great German scientists and Nobel Laureates of that time, Max Planck (1858-1947), Ludwig Prandtl (1875-1953), Albert Betz (1895-1968), Max von Laue (1879-1960), Otto Hahn (1879-1968), Werner Heisenberg (1901-1976), and my friend from the Ulm High School, Otto Haxel (1909-). The nuclear scientists had returned from their internment in England; their conversation invariably centered on the peaceful use of the atom.

Another Allied activity was the Fiat Review of German Science, under the chairmanship of Max von Laue. Covering the period from 1939 to 1946, it included all physical sciences and medicine. Accepting an invitation I was happy to contribute the section "Verbrennungsvorgänge" (combustion phenomena) for Volume VIII "Hydro- und Aerodynamik," published by Albert Betz.

Appendix 4.1

The Gas Turbine as a Vehicle Powerplant

1946 Monograph

Contents:
1. Introduction
2. The Components
   a. The Compressor
      Axial-flow
      Radial-flow
      Combination
      Lysholm
      Vibrations
   b. Regenerator and Combustor
   c. Turbine
   d. Gears and Bearings
3. The Powerplant
4. Other Processes
   a. Vacuum process
   b. Thermal compression
5. Summary and Conclusions

Noteworthy items:
Change of specific power and efficiency with size.
Regeneration, water injection, multifuel.
Effective massflow limits.
Screw compressors for low massflows (Lysholm).
Ceramic materials.
Low-cost turbine blades.
Concept of vehicular gas turbine.

Our horizon is never quite at our elbow.
Henry David Thoreau (1817-1862)

Chapter 5

The Automotive Gas Turbine

Journey to England

Early in 1947 the compilation of German scientific and engineering reports and monographs by the Allied occupying authorities came to a close. Because of my contributions, I received a contract offer from the British Ministry of Supply (MoS) to work as Research and Development Engineer in England, initially for six months, starting as soon as the formalities with the British and American authorities were completed. I knew of various groups of German engineers and scientists working in England. Although I had no official assurance, I assumed I would join the gas turbine team in Newcastle-on-Tyne.

The decision to accept this offer presented a bold step into the unknown, both professionally and personally; it involved leaving my wife and three young children on their own for almost a year and parting with many friends. In keeping with the adage, "One's standpoint depends on where one sits," my resolve was judged in many different ways, from those who appreciated it as a great professional opportunity, to those who decried it as a mild form of treason. Leaving the fatherland in its time of need was incomprehensible to a few.

In late Summer 1947 I departed from our home in Bissingen, first with an escort in a British jeep to Völkenrode, then alone by train to Hoek van Holland, the Channel Ferry to Harwich, by boat train to London, and finally, for a few days to the complex in Wimbledon, where German scientists and industrialists had been detained for questioning after WW II. While there I went for a walk. I took great care to remember where I was going so that I would find my way back again without having to ask someone for directions. I was afraid that my status as an enemy alien and my Teutonic accent would cause an undesirable street scene. My concern was, in fact, overly cautious. During my life in the United Kingdom, not one incident occurred arising from my being a former enemy. On the contrary, people were often anxious to learn what life was like on the other side. On reaching the main street, I heard in the distance a marching band, reminding me of the many victory parades I had seen in the French zone. I was wrong! It was the Salvation Army collecting contributions for its worthy cause.

After my short stay in Wimbledon, I was put on a train for my final destination in Northumberland. The German contingent lived in Whitley Bay, a seaside resort. They welcomed me with a question, "What took you so long?" They then informed me about our job and its origin.

Sir Claude D. Gibb (1898-1959), Chairman and Managing Director of C.A. Parsons & Co., Newcastle-on-Tyne, had been Chairman of the Ministry of Supply Tank Board during the war, and Dr. Andrew T. Bowden (1897-1969) had been a tank designer until 1945. After the war, they had gained access to German tank gas turbine developments. Also, Parsons operated a gas turbine laboratory and its marine division worked on gas turbines for ship propulsion. Thus, the MoS chose Parsons for the design and development of a gas turbine of 1000-horsepower output for a battle tank.

Bowden was in charge of this classified project. He had assembled a team whose members had come from Germany: Paul Kolb (primus inter pares), Dr. Waldemar Hryniszak (aerodynamics and thermodynamics), Ludolf Ritz and Paul Hentrich (heat exchangers) and Otto Zadnik (transmissions). I complemented the team with my specialties: component development, turbine cooling and vibrations. The group was ethnically well-balanced: two Swabians, two Austrians and two Prussians. A British draftsman and a tracer completed our gas turbine design department, located in a loft of the Parsons office building.

No restrictions were imposed on our connections to the various Parsons departments, shops and facilities. Initially our personal contacts were almost exclusively with the Materials Department. The other department heads kept their distance and considered us more as strangers than as colleagues. Other factors contributed to this contrast. Charles A. Parsons (1854-1931) had founded the company in 1889. He was honored with a knighthood, a Memorial Lecture series, a Memorial Library and, in 1950, with a Memorial Window in Westminster Abbey. Thus, Parsons was an old, venerable company with a great tradition; its products dominated the steam turbine and electric powerplant market. We could offer only our brain power to advance gas turbine technology. The distinction between the steam turbine, industrial gas turbine, and vehicle gas turbine people was also reflected in the measurement systems they used. Steam traditionally preferred the British system with degree Fahrenheit and British thermal unit (BTU); industrial gas turbine used a mixture of British and metric units, e.g., temperature in centigrade and centigrade heat unit (CHU). We, of course, used the metric system with centigrade and kilocalorie. For manufacturing we converted our drawings to British units.

The Project Office of the MoS's Fighting Vehicles Research and Design Establishment (FVRDE), located in Chertsey, Surrey, south of London, monitored our work and progress with frequent phone calls and reciprocal visits.

The Tank Gas Turbine

The design of the tank gas turbine which was on the drawing board incorporated a gas generator, consisting of a centrifugal compressor driven by a single-stage axial turbine, two rotary regenerators and four combustion chambers of the Whittle-type. A two-stage axial turbine delivered power via a reduction gear to the transmission, which was designed for the high torque characteristic of the power turbine and also for the tank's differential steering requirement.

Under normal operation the power turbine was mechanically independent of the gas generator. A Sinclair synchrocoupling connected the power turbine to the gas generator shaft whenever the power turbine speed tended to become relatively higher than the gas generator speed. That novel arrangement served two purposes: (1) it prevented dangerous overspeeding of the power turbine and (2) it utilized the compressor power for braking the vehicle.

The first conceptual layout of the engine was based on the then-high turbine inlet temperature of 1000°C (1832°F), highly effective rotary regenerators and good efficiencies of the turbomachines. Prognosis was for low specific fuel consumption to be equal to that of diesel engines. The detailed design, however, revealed various pressure, leakage and heat losses which increased the fuel consumption figures to those of Otto engines.

On completing my studies, I agreed with the aerodynamics and thermodynamics of the engine layout but was concerned about some mechanical design configurations. For example, the connection of the gas generator and the power rotors through the hollow power turbine shaft with a thin quillshaft appeared far too risky. An analysis of the drive system for torsional resonances was indicated; some other vibratory problems also had to be addressed.

For the ensuing analytical and design work, we came in on a Saturday. Due to the classified nature of our project, our design office was under the strict security of the company guard corps who opened and closed it on weekdays for the regular work hours. The guardsman on duty, somewhat surprised, handed us the key. It carried a large label, "German Tank Team." We were amused and flattered, too.

Because all of us were newcomers to tanks in general and their engines in particular, we suggested in-situ familiarization at FVRDE. That conference was conducted in two parts. A Royal Army general presented the major requirements for a better tank powerplant. He pointed out that present engines require a gear shift every 60 seconds, a heavy burden on the driver on the battle field. Cognizant of the superior torque characteristic and other advantages of the gas turbine, he expressed the hope and wish that we would dramatically advance the state-of-the-art of tank propulsion. Finally, he emphasized the crucial factor of reliability. All advancements amount to naught if the tank loses its power source, he said. We all nodded in agreement, indicating that we got the message loud and clear.

A Centurion tank was made available for a demonstration in the field. Standing on the tank next to the turret, I held on to the gun barrel for my safety. I soon noticed that a 50-ton mass does not accelerate fast in any direction, not even on terrain full of craters. Therefore, I felt comfortable standing freely and counterbalancing the tank motions. After criss-crossing the field, we drove on a paved road at full speed, which apparently was only moderate; a youngster on a bicycle had fun passing us.

Stopping on a hill, we had the landscape explained to us. There the faint smog over London, there the clear sky over the English Channel. "Fine," the general said, "let's have lunch." Inadvertently he proved his point of dependability; the Centurion tank engine would not start, and some of us had to walk back to the Establishment!

In addition to the cooperation of FVRDE, we enjoyed the assistance of De Havilland Company for the radial compressor and of Joseph Lucas Co. for the combustion system. Gradually our group was expanded, with E.P. Hawthorne[1] (1920-) as formal head and a few development engineers with gas turbine or related experience: Charles S. Lowthian, William A. Russell, and Matthew Potts.

Representing a congenial team, with expertise on one side and eagerness to learn on the other, we enjoyed excellent cooperation with our British colleagues. Our rather imperfect English language seldom posed problems. We also found that good stories and jokes travel across language barriers.

Our engine project progressed steadily but slowly. Delivery of experimental parts often missed the given date, slowed the pace and threw our scheduling into disarray, a frustrating experience to us. Our English friends advised us as to how to eliminate these worries by using their formula for parts deliveries, "Multiply the promised length of time by two and add a fortnight!"

Still, our work was appreciated and my contract was extended. During my vacation in Germany, my wife and I decided on the family's move to England. For familiarization and acclimation, all of us first lived with my landlord and his wife in Whitley Bay. After overcoming the initial shock of a foreign language, our children liked their school and their teachers; they settled nicely. We wanted to acquire our own home, though, despite the financial difficulties involved. Since the West German currency reform had practically wiped out our savings, we needed a

---

[1] Edward's older brother, Professor Sir William R. Hawthorne (1913-), had the honor to represent the United Kingdom at the DGLR Symposium "50 Years of Jet-Powered Flight" in Munich. He presented a paper entitled "The Early History of the Aircraft Gas Turbine in Britain." He also brought greetings from his brother; Magda and I enjoyed our pleasant conversations with him about our time in England.

first and a second mortgage. The interest rates frightened us but we had no choice and took the plunge with a semidetached house in nearby Monkseaton.

The Government checked our whereabouts annually. A policeman came to the door and announced this as his duty. When my wife invited him in for a cup of tea, he gladly accepted with the words, "I am a foreigner, too. I'm from Scotland!" This remark made us feel at home in the land of the friendly Geordies (as inhabitants of Northumberland are called).

Professionally, the potential vibration problems of the drivetrain bothered me.[2] Damping probably reduced the danger level of resonance conditions when driving on muddy grounds but not on hard surfaces. However, I disliked a design with no leeway for change. To eliminate any critical speed that might cause a breakdown, I replaced the quill shaft with a geartrain to the gas generator rotor. Also, all high-speed gearing was redesigned with single- and double-helical teeth. These modifications increased the engine weight but also raised my confidence in the integrity of the powerplant. Voysey, now a Ministry official, agreed with me, stating in a memorandum, "Bentele has done all he can do within the framework of the present design to avoid trouble."

In contrast to aircraft propulsion engines, vehicular gas turbines operate over a wide range of power and speed. The fuel consumption should be low for all driving conditions. This requirement can be achieved with effective heat exchange, where the exhaust gas energy is transferred to the compressed air. Ritz had applied the Ljungstrom rotary regenerator to gas turbines at AVA in Göttingen and finally to this tank engine. Rotary regenerators offer considerable advantages over stationary recuperators: much higher efficiencies, smaller size and lower weight and cost. Their Achilles heel, however, is potential leakage of the precious compressed air to the exhaust. If this loss exceeds a few percent of the working air flow, their effectiveness drops to that of a recuperator.

The drum-type rotary regenerator selected for the tank engine is schematically shown in Fig. 5.1. The drum is peripherally divided into eighteen departments and axially into six sections. The chambers contain the matrix, at that time usually a wire-mesh package. On the low-pressure side, exhaust gas from the power turbine stores its heat in the matrix; on the high-pressure side, the matrix releases that stored heat energy to the compressed air. Sealing shoes are provided to prevent leakage from the high-pressure to the low-pressure side. In addition to this potential leakage, a carry-over loss occurs because a chamber full of compressed air releases that air volume to exhaust.

---

[2] At that time, vibration problems of propeller gas turbines were rampant and had caused many difficulties that eluded satisfactory solutions.

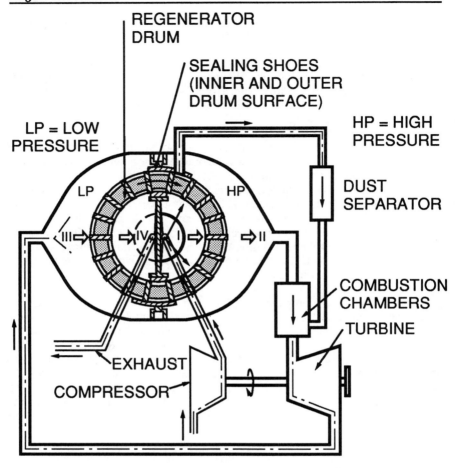

REGENERATOR
DRUM

SEALING SHOES
(INNER AND OUTER
DRUM SURFACE)

LP = LOW
PRESSURE

HP = HIGH
PRESSURE

DUST
SEPARATOR

COMBUSTION
CHAMBERS

TURBINE

EXHAUST

COMPRESSOR

Fig. 5.1. Parsons rotary regenerator (1947).

A variety of factors determines the effectiveness of such a regenerator: the characteristic of the matrix (heat storage, flow conditions, and corresponding pressure drop, volume of metal and air in the chamber, etc.); rotational speed of the drum; leakage and carry-over loss; mechanical friction of the seals; and driving power of the drum. Hryniszak tried to arrive at the proper matrix design and flow conditions with long, complicated mathematical formulae that contained all relevant aerodynamic and thermodynamic numbers. Ritz and Hentrich also were busy trying to find the proper heat exchanger configuration and matrix. In my opinion, all these optimizations were overshadowed by the leakage losses. I concentrated on these and on the proper operation of the seals. For this purpose, we introduced pressure compensation of the sealing shoes to reduce the driving power and the leakage rate. Later on, I designed and experimented with lubrication systems with solid lubricants that were suitable for hot metal surfaces such as colloidal graphite and molybdenum disulfide. Expected clogging of the matrix air and gas passages was eliminated by continuous cleaning with a burst of air as indicated in Fig. 5.1. British and German patents were granted for these devices.

Once I had to attend a meeting alone at FVRDE. Two incidents during this trip stayed in my memory. Passing the courtyard of the Establishment, I noticed a variety of captured German tanks and armored vehicles, the standard Panthers, Tigers, and, as a unique monument, Porsche's 90-ton monster, the Maus. In the guard room, a veteran whose uniform was adorned with ribbons and medals from the Boer and First World War asked me to sign in, and I did. When he looked at my entry, he hesitated, turned pale, looked again, and said, "Sir, I see in the column 'Nationality' you wrote German. I sure hope it's all right." I replied, "Yes, I think it will be." Both of us were relieved. My conference with the FVRDE engineers was equally harmonious and successful.

On my return trip to Newcastle, the six-seat compartment of the night train was fully occupied. The Englishmen, some of whom had served with the occupation authorities in the British zone, told their stories, boasting how they sometimes let the Germans have it. In Hull all but one left the train. He said to me, "You have not taken part in our lively conversation." With my reply, "No, I haven't," I revealed myself as a foreigner. In this case, as in many others, the polite Englishman then asked me where I was from. I always replied, "Take a guess." Looking at my dark blond hair, the questioner invariably started with the Scandinavian countries, Norway, Finland, Sweden and Denmark. When each time I said no, he jumped to Switzerland. "You missed Germany, didn't you," I said. He almost apologized and said, "Oh my, our conversation must have really annoyed you." I said, "No, not at all. Nothing was said that I had not heard or read before." With this reassurance, we had a most pleasant chat until we reached Newcastle.

In 1950, the Rover Company of Sollihull, Birmingham, publicly demonstrated a gas-turbine-driven passenger car on an abandoned airfield. This "first" provided a boost for the development of vehicular gas turbines, more in psychological than in technological terms. Figure 5.2 depicts the car and the event with the official certificate of the Royal Automobile Club. The gas turbine installation in the vehicle failed to convey the size and weight advantages of a gas turbine; there was only one passenger seat next to the driver, the rest of the car was filled with the gas turbine machinery, combustor, and intake and exhaust ducts. Critics of passenger-car gas turbines liked the car's license plate, "JET 1": the proverbially high fuel consumption of jet engines would indicate the car's low fuel mileage. The performance of engine and car was also far from superior. It took 13.2 seconds to start the engine and 3.4 seconds to move the car. Its acceleration was only moderate: it took 14 seconds to reach 60 mph. Although being far from spectacular, this demonstration was considered the "Kitty Hawk" for automotive propulsion. Many automobile companies pursued gas turbine development, some in earnest, some as a status symbol.

In 1951 I had the opportunity to visit the Automobile Show in London's Earls Court. There I had a most illuminating conversation with a sales representative on the Rover stand:

MB: I'm disappointed about your exhibits. You see, I inherited a small fortune from an American uncle (a white lie, of course) and wanted to buy one of your new gas turbine cars, and you don't even have one as an exhibit.

Rep: Well, the car requires some more refinements.

MB: By the way, will it be with or without?

Rep: What do you mean, with or without what?

MB: A heat exchanger. I'm accustomed to long trips and with your demonstrated fuel economy of some 4 mpg I wouldn't get very far, would I?

Rep: You're right. Our engineers are working on the heat exchange.

MB: Fine. Another point. What brakes will the car have?

Rep: I'm amazed at your question, sir. Four-wheel brakes, of course.

MB: That's what I thought. But there's the rub. You probably noticed from my accent, I'm from the Continent. I drive a lot in the Alps. Going down steep and long hills, I put my old jalopy in low gear so that the engine absorbs the gravity power of the vehicle and I hardly need any brakes. Your free power turbine will not be able to do that. I will have to rely on your brakes. These will get hotter and hotter, eventually fade, and I would be heading for a disaster, wouldn't I?

Rep: You have a point there. I will tell my engineers.

MB: You better do.

Rep: Thank you for your interest in our gas turbine car. I hope to sell you one in a few years from now.

MB: You mean in a few decades, if ever.

Now back to our gas turbine. In view of the slow progress with the regenerator, it was decided to test the gas generator without it and simulate the power turbine with a jet nozzle; in other words, explore the powerplant as a jet engine. Knowing that this engine run would become a great affair for Parsons and the Ministry, I kept the final build fairly quiet. When, after hard and frustrating work, the engine was finally installed on the test bench, I said to the mechanic, "Fine, everything is now go. Please connect the starter motor cable to the battery." He looked at me in disbelief and said, "Doc, I cannot do it. I'm not an electrician." I was flabbergasted but then realized that he was strictly following the prevailing rules of his labor union. On my friendly persuasion, he connected the cable, we turned over the rotor, lit the four combustion chambers and ran the engine with self-sustaining and higher speeds.

94

ROYAL AUTOMOBILE CLUB

Report of Test
of

A ROVER GAS TURBINE CAR

Wednesday, 8th March, 1950

(HELD UNDER THE OBSERVATION OF THE ENGINEERING AND TECHNICAL DEPARTMENT OF THE ROYAL AUTOMOBILE CLUB)

The Rover Company Limited, Solihull, Birmingham, England, submitted a car fitted with a Gas Turbine Power Plant for Test.

The following leading particulars apply to this car:—

Wheelbase	9′ 3″	Type of Body ... ... Open
Track	4′ 4″	Number of Seats 2/3

The general design of the car, apart from the power ... conventional lines, and its external appearance normal.

The Test was held on the Motor Ind... circuit at Nuneaton, Warwickshire, and at the tim... wind speed 5 m.p.h., temperature 54 F., Barome...

The Entrant intended th... Ltd. in the application of the Gas Turb...

The power plant, ... centrifugal compressor, dual comb... dent power turbine. The latter ... to drive a conventional rear...

The fuel ...

The meth... button on the instrument pan...

The time ... was 13-1/5 seconds, and the car moved for...

Control ... solely by means of the accelerator pedal, the only ot...

A lever ope... normally.

The car was ... of the Proving Ground measuring approximately 2.75 miles per lap, ... utomobile Club Observers travelled alternatively as drivers and passenge...

No attempt was ... attain maximum speed, but during the course of the Test a speed exceeding 85 m.p.h. was re... attained, at which speed the compressor-turbine revolution counter indicated 35,000 r.p.m.

In a test of acceleration from standstill, the car smoothly attained 60 m.p.h. in 14 seconds.

Although no provision for silencing the exhaust was observable, the volume of noise was not excessive or unpleasant, but was naturally accentuated during acceleration.

Chairman

Fig. 5.2. World's first gas turbine car (1950).

When I reported my successful initial operation to Hawthorne, he took a dim view of it. He had wanted to invite the Parsons and Ministry officials to witness this maiden run. I simply replied, "I knew that, but are you aware of the visitor syndrome? I wanted to be sure everything is OK and avoid the slightest possibility of a mishap for the official demonstration."

For this event, Sir Claude honored us for the first time with his presence in our test facilities. Bowden proudly presented his project to the distinguished audience. Thanks to my previous dry run, the demonstration proceeded with starting, lighting all combustion chambers, and running at various speeds, without a hitch. Snap accelerations impressed those who were used to the slow speed changes of steam and industrial gas turbines. Sir Claude was pleased with Parsons' first jet engine run, and congratulated and thanked us for our good work. There was no champagne dinner, though, as would have been the custom on the Continent.

The diligent design effort of Kolb, Zadnik, our British engineers and me led to a neat power package for an installation and demonstration in a tank hull. Its final version is shown in detail in Figure 5.3.

Parsons and Ministry officials were pleased with our progress and arranged for a Royal visit. In November 1951, Prince Philip, Duke of Edinburgh, toured the Parsons facilities, including our offices and test benches. My wife and I were at the time celebrating her parents' golden wedding anniversary in Germany, so I missed the opportunity to shake hands with a high member of the Royal family.

Thermal Shock Investigations

As already mentioned, gas turbines for land vehicles operate distinctly differently from those for aircraft. The latter employ only a few power regimes: takeoff, climb, cruise, and descent/idle. In contrast, vehicle engines are subject to frequent power and speed fluctuations, from high power for acceleration and maximum vehicle speed, to low power and idle. In gas turbines, these alternating operating conditions cause temperature variations of the combustion chamber and turbine resulting in thermal stresses usually referred to as thermal shocks.

Our tank gas turbine was particularly prone to these detrimental effects, due to its novel feature of utilizing the compressor power for vehicle braking. My major concern was the turbine blading. I started to address this problem systematically. Since my literature search revealed only scant information on the subject, I had to consider the theoretical aspects first, for which my associate Lowthian provided invaluable assistance.

Beginning with static conditions, we calculated the thermal stresses in a slab for the two cases of constant temperature difference and constant heat flow. To obtain an all-encompassing picture we chose four materials with different physical,

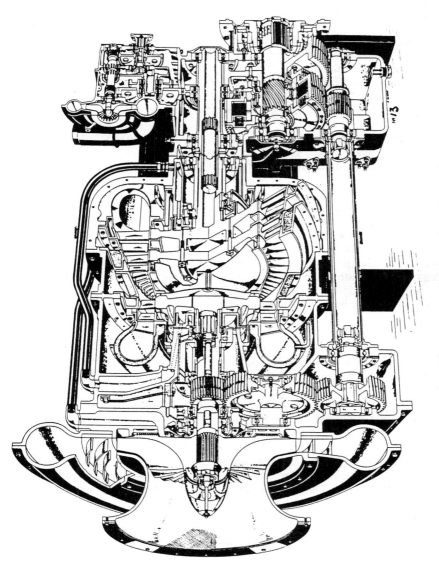

Fig. 5.3. Parsons tank gas turbine (1947–52).

thermal and mechanical properties. As gas turbine material we chose a ferritic high-temperature steel and a high-temperature alloy and as extremes on either side — copper and porcelain. Using the results obtained for static conditions, we evaluated transient conditions whereby the existing temperature distribution of the specimen is suddenly changed on its surface.

Extending these calculations to shapes approaching those of turbine blades, we arrived at a qualitative relationship for the number of thermal cycles to failure. This number increased with material strength, thermal conductivity and diffusivity; it decreased with linear coefficient of expansion, density, specific heat, modulus of elasticity (Young's modulus), cross-section, change in cross-section, local heat transfer coefficient, temperature and temperature gradient.

We had investigated theoretically the effects of these factors individually, but the relative importance of any one of them was exceedingly difficult to estimate. Considerable experimental results were necessary to obtain quantitative data.

Published thermal shock tests were haphazard and too far away from the real world. Our target for the number of cycles to failure was in the region of thousands; the available tests ranged from less than 10 to less than 100, and only two cases with 360 and 364 cycles were reported. These low numbers were apparently obtained by "accelerated testing," i.e., more severe shocks than those occurring in practice. I was aware that any estimation of long-term properties relating to creep and vibratory fatigue strengths from static or short-term tests was apt to be very misleading. So it appeared to me the same would apply to thermal shocks. Thus, I drew the analogy between thermal fatigue and vibration phenomena.

Applying this reasoning, we verified our theoretical thermal shock investigations in rig tests with actual first turbine nozzles of our tank engine. Thereby, we also found evidence that the surface conditions of the material have similar effects in cyclic thermal as they have in cyclic load applications.

Summarizing our work, two test rigs for thermal shock tests on gas turbine materials were proposed, one simulating turbine blading and one combustion chamber walls.

The reputable British magazine *Aircraft Engineering* published our pioneering work in an article entitled "Thermal Shock Tests on Gas Turbine Materials" in its February 1952 issue. The editorial appreciated our fundamental treatise of the subject, the analogy between thermal shock and vibration phenomena, and the proposed test rigs and test conditions. It was the sole published paper that was directly related to our tank engine, although it did not name it. The editor and Lowthian and I, the authors, were justly proud of it.

Control System

My first visit to FVRDE and the Centurion tank demonstration made it clear that the full potential of a gas turbine powerplant can be realized only if it contributes no additional burden on the driver. Consequently, the control system has to provide for maximum vehicle acceleration and deceleration within the allowable mechanical and thermal conditions of the gas generator and power turbine under all ambient conditions. Also, starting and vehicle braking, which pose the danger of flame extinction in the combustion chambers, have to be included.

I conceived and schematically designed a purely mechanical system with cams, levers, fulcrums and so on. It fulfilled all requirements. Compared to modern electronic systems it looks primitive, but it was novel at the time and served its purpose well. A British patent applied for in 1951 and granted in 1955 testifies to this effect.

Industrial Gas Turbines and Other Events

The gas turbine engine had revolutionized aviation. It was also considered as a new powerplant for a range of other, less dramatic applications. After WW II, Great Britain's sole source of heat and power energy was coal. Sir Claude, honored to deliver the Watt Anniversary Lecture for 1949, chose the title "Britain's Energy - A New Conception." In it he made a variety of proposals for the effective utilization of coal. In a number of technical papers, Bowden also had expressed the view that gas turbines in industrial applications have to be capable of burning coal.

The experimental machine in Parsons' gas turbine laboratory had been operated with a variety of low-cost liquid fuels, and much experience on corrosion and deposits in the hot gas turbine sections had been gained. It appeared natural to run this gas turbine with coal and to be "first." To achieve this target, the available combustor was modified only for the introduction of pulverized coal. All provisions to minimize the detrimental effects of the coal ashes on the combustor and turbine components were postponed.

Pulverized coal was stored in a large cylindrical/conical hopper from where it was mechanically fed into the combustion chamber. For expediency and to assure continuous flow of coal powder to the feed mechanism, the hopper cone was fitted with a pair of scraper blades and handles. These were slowly turned by two men on a walkway around the hopper. Though not very "high tech," this arrangement served the purpose well.

The inaugural gas turbine operation with coal was attended by officials of the Ministry of Fuel and Power, the Ministry of Supply, and other government agencies. A schematic drawing and chart depicted the major gas turbine components. The ongoing procedure was continually explained: First, lighting and running solely with fuel oil, then gradually introducing coal at an increasing rate and simultaneously reducing the fuel oil injection, finally shutting off the oil, and the gas turbine running solely on coal. The audience applauded this feat and took notice of the smooth operation of the gas turbine for an appreciable length of time. Fortunately (or deliberately), the exhaust stack with a billowing black smoke was behind the building, out of sight. An MoS official, associated with our tank project, took me aside and said, "Great, it was proven that coal is still burning." Newspapers reported the event and Parsons pamphlets and ads in trade magazines marked the historic achievement.

The tremendous efforts in Great Britain and the United States to develop coal-burning gas turbines for locomotives and industrial applications stumbled on the flyash and corrosion problems. To the best of my knowledge, no coal-burning gas turbine of this period ever reached the production stage. Recently, coal combustion in fluidized beds under atmospheric or elevated pressures (AFBC or PFBC) appears to be working, not only for steam boilers but also for gas turbines.

In 1981, the General Motors Corporation celebrated the 25th anniversary of its Technical Center. To emphasize the fact that there is more coal in the U.S. than petroleum, gas turbines burning powdered coal were demonstrated in an Oldsmobile and a Cadillac. It was pointed out, however, that no production was planned for such powerplants. It was also conceded that nearly everything conceivable would pose a problem. Remembering the Parsons coal-burning gas turbine run of 30 years ago, I thought, "Plus ça change!"

As an expert on air-cooled turbines, I exchanged visits with Parsons marine division. It explored water-cooling of gas turbines, which had been experimentally investigated in Germany. In theory, two methods of water-cooling are feasible, internal cooling with a recycling waterflow and external cooling with water spray onto the blading.[3]

Internal water-cooling of hollow blades increases the heat transfer coefficients and substantially reduces the blade temperatures. High turbine inlet temperatures had been demonstrated with blades made from ordinary steel and even aluminum. The high centrifugal forces, however, cause stress corrosion. One pinhole in a turbine blade suffices to empty the water cycle and ruin the turbine. External water-cooling also suffers from detrimental effects of corrosion and deposits.

I am unaware whether these water-cooled turbines made it to production. The cooling effectiveness of water is, however, so alluring and attractive that this method is revived every decade or two. In 1984 it started to make waves again with applications in industrial turbines.

---

[3] In Whittle's first experimental jet engine, the turbine disk was kept cool by a shield with circulating water.

One event finally brought our three departments closer together. Sir Claude had secured a contract for two powerplants of 100,000 kilowatts each for Toronto, Canada. They were planned to operate on weekdays and to be shut down for the weekend. Parsons' biggest units then were in the 60,000 kW class. For expediency, these were apparently photographically scaled to 100,000 kW.

After a relatively short period of operation, one of the units experienced a catastrophic failure. The conference in Toronto, where Parsons' engineers debated the causes of the disaster with their Canadian customers, received another shock: The second unit suffered the same fate as the first one.

Nothing as drastic as that had ever happened to Parsons before. So we "outsiders" also were approached to assist in finding out what had gone wrong. When all returns were in, two factors finally emerged as the culprits: (1) steam powerplants are normally running for long periods of time without shutdowns; for the Toronto 100,000 kW size the operating schedule was entirely new and had caused a low-cycle fatigue failure; (2) scaling of the powerplant exacerbated thermal expansion effects which proved noncritical in the 60,000 kW units but caused problems in the larger ones.

The lessons learned from this dismal experience were twofold: (1) The operating schedule of a powerplant drastically affects its useful life. We knew that rule from our aircraft engines. In test runs simulating fighter or bomber operations, the fighter schedule always produced component failures first. (2) Substantial upscaling of a powerplant has to be accompanied by a strict scrutiny to determine whether marginal design features of the original are apt to become major deficiencies in the scaled version.

Our contribution was gratefully appreciated, and our status in the Parsons organization greatly enhanced.

The Toronto story parallels that of Tacoma's disaster with the Galloping Gertie bridge. Caveat to engineers who rely solely on parochial tradition: They may one day face disappointment at best and humiliation at worst.

In addition to our work on vehicular and industrial gas turbines, we tried to keep abreast of jet engine developments with visits to the Flying Display and Exhibition of the Society of British Aircraft Constructors held at Farnborough, Hants. We studied the advancements in the gas turbine field and enjoyed the newest aircraft.

Spectacular were prototypes of the Princess flying boat and the Bristol Brabazon, then the largest passenger airplane. It was conceived and designed in the latter years of WW II by a committee headed by Lord Brabazon (who carried the British Flying License #1) to establish a British foothold in the coming lucrative transatlantic passenger market.

The Brabazon exceeded the American passenger planes Lockheed Constellation and Boeing Stratocruiser in wing span and in propulsion power by factors of close to two. Its eight engines were combined to four propulsion units à la Heinkel's He 177. On its demonstration flights over the British Isles, it once flew over our Newcastle plant, offering a beautiful sight. It always had to return to Bristol's own Filton airport, the only one capable of accommodating it for takeoff and landing. The Brabazon's range, therefore, was zero. The project finally faded away.

Another timely event brought the British Atom Train, a traveling exhibition on atomic energy. I remembered my conversation on the subject with the German nuclear scientists at AVA Göttingen and the exhibit improved my knowledge considerably. It demonstrated objectively in pictures and graphs the fundamental facts of the atom and its practical applications: the first release of atomic energy, the atomic bomb, and on the brighter side, atomic power generation, atomic energy in medicine, agriculture and industry. The exhibit showed vast possibilities of atomic energy for good and evil. It concluded with the question, "What is it to be?" At present, mankind is further away from an answer to that question than 40 years ago.

During my stay in Germany in November 1951, Ernst Heinkel persuaded me to rejoin him in the Stuttgart plant as Chief of Development. The tank engine project was moving along, but a final development for production of the engine appeared out of the question. Parsons' gas turbine development for industrial and marine applications was at the fringes of my interest and capabilities. With these hardly attractive prospects in mind, I asked for release from my employment contract and permission to return to Germany. My request was graciously granted.

I was grateful for the confidence extended to me and the appreciation of my work on the tank engine. Bowden expressed it in a certificate as follows: "His work has been of the highest technical nature and he has brought to the job unique experience and ability of which I can only speak in the highest terms."

My colleagues honored me in a fine ceremony with a nostalgic present, a watercolor of St. Mary's Island. My family often had enjoyed pleasant visits to this beautiful landmark off the Whitley Bay coast.

In April 1952, the Bentele family bid goodbye to its friends with an "Auf Wiedersehen!" All of us had adjusted well to the English customs and lifestyle; we especially appreciated the kindness and friendliness of the Geordies. Our stay in the United Kingdom was a great experience none of us would have wanted to miss. After some 40 years, we still reminisce about our time there and relive it with occasional visits.

Postscript: In 1954 the Ministry of Supply and the Society of Motor Manufacturers and Traders gave a demonstration of British Military vehicles at FVRDE before several hundred visitors, many of them from overseas. The appearance of a tracked vehicle powered by the MoS/Parsons tank gas turbine engine was the outstanding feature of the event. It represented the first stage in the practical investigation of gas turbines for tracked vehicles. This mobile test bed demonstrated the advantages of gas turbines for tanks: their excellent torque characteristics, low weight and small size, low noise and no smoke, ease of starting, as well as power absorption for braking the vehicle. It also revealed the necessity for high efficiencies of the turbomachines and the heat exchanger to lower the fuel consumption to acceptable levels.

The engine used for this demonstration is now on permanent display at the Tank Museum at Bovington, Dorset, as a symbol of the pioneering work performed at Parsons in the late 1940s and early 1950s.

The U.S. Army issued an RFQ for a tank gas turbine in the mid-1960s. I participated in the competition at Curtiss-Wright and in the development at AVCO Lycoming. It was sad for me to notice that our work at Parsons was completely ignored, unknowingly or deliberately, and that the U.S. Army's development program was not presented as a revival but as something entirely new!

A man of genius makes no mistakes.
His errors are volitional and are the portals of discovery.
James Joyce (1882-1941)

Chapter 6

Stuttgart Again

The New Heinkel Company

Right after the 1952 Easter holidays, I started work in the Heinkel plant in Stuttgart-Zuffenhausen, which I had left in 1946. It was quite a change from my sojourn in England, in more than one respect. There it was one product, a tank gas turbine, one customer, the Ministry of Supply, and one R&D team. Now I was confronted with a multitude of products, one major and thousands of other customers, a fairly loose organization to accomplish all facets, from conception of the product through development and manufacturing to marketing and sales to the general public, without government sponsorship.

Fortunately, I found some of my former Heinkel-Hirth associates, design, test, development engineers and technicians, who had welcomed me with open arms on my return from England. We formed a team of mutual trust, assistance and cooperation ready to face any challenge.

Ernst Heinkel had obtained repossession in 1950 of the Heinkel-Hirth plant in Zuffenhausen.[1] All his other facilities, located east of the Iron Curtain, had been confiscated, their machinery and equipment shipped to the Soviet Union or destroyed. Finally, he even lost his share of the Jenbach plant in the Austrian Tyrol. To raise working capital for his new commercial venture, he had sold a portion of the previous Heinkel-Hirth plant including the employee amenities, cafeteria, sporting field, and executive dining room. He had complemented the remaining Heinkel-Hirth group with his aviation cronies and newcomers from all corners of the European continent. It was quite a conglomerate of individualists and prima

---

[1] The Hirth family earlier had started, with the help of Curt Schif, a factory in nearby Benningen/Neckar for the development and manufacture of Hirth engines for the agricultural and industrial market. To prepare for this venture, engine concepts and designs had been worked on toward the end of the war. The Hirth engines were again highly successful. Unfortunately, the company extended itself during the snowmobile boom years. The collapse of this domestic and North American market took, I believe, the Hirth company with it.

105

donnas, who tried to prevail with their opinions. However, all knew there was only one man who called all the shots — Professor Dr.-Ing. E. h., Dr. phil. h. c. Ernst Heinkel.[2]

Headquarters was his villa Robert Bosch-Strasse 3 on a hill overlooking Stuttgart, with a garden in front and rear including a mini-mini swimming pool. The first floor was the office for his various enterprises. He resided with his wife, Lisa, and their son, Karl Ernst, on the upper floors.

My duties as Chief of Engineering Development comprised design, test, service and repair. I soon realized that Heinkel trusted me with a host of extracurricular activities such as consulting on his gas turbine, aviation and space ventures, his negotiating with domestic and foreign business people, accompanying him on important visits, particularly abroad, representing him at conferences and symposia, and so on. In the best of circumstances, I felt as his alter ego, in the worst, I appeared to be his Girl Friday.

To streamline my organization, I insisted on locating all departments under my jurisdiction (Design, Experimental Assembly & Test, Customer Service & Repair) close together, within walking distance. This ease of personal communication proved highly successful; large automotive companies also have adopted it.

OEM Engines

On my arrival, the Heinkel company had only one major customer, Vidal & Sohn Tempo Werk, Hamburg-Harburg. The Vidals, father and son, had pioneered light vehicles for the transportation of goods and passengers. They carried the tradename "Tempo Matador," followed by a number which indicated the payload in kilograms. All employed front-wheel drive, two-stroke engines up to 1-ton payload and above that four-strokes. Heinkel then manufactured and delivered two-stroke two-cylinders of 400 and 450 cc with 15 and 17 horsepower, and a four-stroke four-cylinder 1100 cc with 34 horsepower. The specific power of 40 hp/liter of the two-strokes and 30 of the four-stroke was fairly respectable.

The two-stroke engines were based on an outside design of extreme simplicity using the Schnürle loop-scavenging scheme and the then-common fuel/oil mixture. The intake and exhaust ports as well as the scavenging conducts and ports were finish-cast in the iron crankcase-cylinder casting, and the cylinder head was cast aluminum. Except for the carburetor throttle, there were no moving parts in the airstream. The crankshaft was built up with shrink fits of the individual components, the main bearings were all of the antifriction type, rollers for the main and

---

[2] Honorary degrees of professorship, Doctor of Engineering and Doctor of Philosophy.

connecting rod, needles for the piston pin and one ball bearing for the axial location and thrust load. A high degree of commonality of the engine components assured low cost in production, maintenance and service.

German law required certification of the engines by the Federal Office for Motor Vehicles in Flensburg, Schleswig-Holstein. Government representatives supervised acceptance runs on our test stands for engine performance, cold starting, as well as a one-hour durability test. All test equipment and instrumentation were sealed before each test series and inspected after it with typical German thoroughness.[3] After completion of all paperwork, the Office issued a Type Certificate and a registration number which had to be included in the engine nameplate. The published technical data had to conform with the results of the acceptance tests. Receipt of the certificate and registration number was always a joyous event.

The service record of our engines was satisfactory except when the engines were operated with insufficient air filtration, with unsuitable lubricating oils or, of course, when they were abused.

One example illustrates the last point as well as the discrepancies between the recommendations of our Operator's Manual — the mandatory ones printed in red -- and the handling by the user and sometimes the dealer. On a tour in the Black Forest, I had the following conversation with a Tempo dealer. (I): "You like the Matadors?" (He): "Sure, they're terrific. I sell many!" "How do you convince a prospective customer of the quality of a 3/4-ton van?" "I take it from the showroom and load it with one ton. See the long steep hill there? I then tell the man: I'll drive it up there and if you're not satisfied with its performance, it will be yours, free!" This dealer apparently had never read the manual: He violated the running-in period and the weight limit, exceeding it by 33 percent. He sold scores of vans, though!

Apart from the expected wear and tear of the engine components, one caused more than the tolerable amount of difficulties. The cage of the roller bearings, a precision aluminum die casting with a minimum of machining, smeared and/or broke too often. I traced both problems to marginal lubrication of the interfaces and improved it with Michelle-type indentations, first machined and later on cast-in.[4]

The four-stroke four-cylinder engine, installed in the Tempo Matador 1400, was designed in-house with emphasis on low fuel consumption. It incorporated a hemispherical combustion chamber with two inclined, pushrod-operated overhead

---

[3] Some inspectors were overzealous. We kidded them with the warning they had forgotten to check the specific gravity of the mercury in the U-tubes.

[4] At Curtiss-Wright, I successfully applied this remedy to the turbine bearing of the turbojet J 65 Sapphire. It operated with metered lubrication and had suffered from the same malady.

valves. Great development efforts were expended to reduce the engine oil consumption to acceptable levels. This problem was fairly common in the German engine and automotive industry at the time.[5]

Manufacturing costs posed a problem particularly in view of the inexpensive two-stroke engines. Cost reduction items often worked well on the test stand but not in the field. An example was the replacement of steel-backed bronze bearings with aluminum bearings that were a lot cheaper. The latter were very sensitive to impurities in the lubricating oil. Reluctantly and with the penalty of a tainted reputation, we had to go back to the more expensive but reliable bronze bearing.

The camshaft drive presented another nuisance. Its large gear contained a fiber-reinforced plastic rim with helical gear teeth whose reliability was sporadic. Exploring alternative drive systems, I sought advice from a friend at Daimler-Benz. Sympathetic to my investigation he said: "You face a technical and a customer problem. If one of our engines, or that of any other well-established manufacturer, experiences such a failure, the customer leniently shrugs his shoulder and says, well, such things can happen. With a newcomer in this field, the customer will be angry and state, look, Heinkel can't even make a simple gear!" Apparently, that was life in the motor industry.

As time went on, we improved our OEM engines and raised their power output with larger bores or by adding a cylinder. Such a three-cylinder two-stroke engine finally found installation in a passenger car, one of the early SAABs of Sweden. Due to some of the drawbacks of two-strokes and the oncoming regulations on exhaust gas emissions, this application gave way to four-stroke engines in the late 1950s; so did all other makes.

Recently, two-stroke engines were and still are heralded again in view of their high specific power and low manufacturing costs. The new requirements of low exhaust emissions and high fuel economy will make them different from those of the 1950s.

Two- and Three-Wheeled Passenger Vehicles

Having essentially only one or two OEM customers is a high-risk business. Therefore, Heinkel wanted to enter the market for personal transportation with his own products. In the early 1950s, a kaleidoscope of vehicles was available for customers with any wish, taste and pocketbook, from auxiliary engines attached to ordinary bicycles, motorized bicycles (Mopeds), motor and cabin scooters, small

---

[5] One example: Daimler-Benz had once been unable to achieve low oil consumption of its top model, the Mercedes 300, on time for the Frankfurt Auto Show. Only an experienced factory chauffeur was allowed to demonstrate the car; otherwise it might have become invisible for the smoke emanating from the exhaust pipe.

and medium-sized cars with wooden, plastic or metal bodies, such as the *Kommisbrot* (Army loaf-of-bread) and *Pappendeckelbomber* (cardboard bomber), as well as orthodox full-sized passenger cars.

As one of his major entries into this crowded field, Heinkel considered a touring scooter. On my arrival, a few experimental machines were being tested. Figure 6.1 depicts the side view of an early production version. The body shape was excellent for the aerodynamics and the comfort of the driver and passenger. For convenience, a small luggage compartment was provided under the seat, as well as a luggage holder at the rear above the spare wheel and a second one in the front, a time clock and, finally, a hook for an *Aktentasche* (briefcase), then the status symbol of the German white-collar worker. The scooter's major attraction was a powerful four-stroke engine in contrast to the two-strokes of all other makes on the European market.

Conceptually, the scooter presented a joy to look at but the hardware required urgent work. Its frame was inadequate. The welds of the intersection of the single front tube and the two smaller rear tubes cracked and broke frequently. My rudimentary stress analysis and fatigue tests conducted in cooperation with the Stuttgart University led to a plate-reinforced welded joint which withstood test drives on rough terrains and achieved a crack-free service record. The engine problems with power, reliability and durability also were resolved. The 150 cc engine produced 7.1 horsepower at 5200 rpm. Top speed of the fully-loaded scooter was 56 mph, guaranteed fuel economy 115 miles per gallon at two-thirds top speed.

After some prototypes performed satisfactorily in rigorous driving on all streets and terrains, the scooter was deemed ready for series production. Its certification included three categories: performance (power, speed and fuel economy), reliability, and safety (lighting, brakes, stability). We passed the tests and inspection with flying colors.

Our marketing people were convinced that the market would accept only a scooter with a retail price under 1500 Deutschmarks (at that time approximately $350). Purchasing and Manufacturing saw no way to produce the scooter for this price. After heated discussions on the pricing strategies, Heinkel himself solved the deadlock, saying "I don't want to lose money with the scooter. We only wish to sell a few scooters per day to a discriminating clientele who will appreciate its exquisite features and pay a price for them higher than the imaginary limit of our marketers. The price will be DM 1650." The shop people were delighted, the head of Marketing, aghast, said, "Well, it's his money. Let's do it and see!"

Simultaneously with putting the dealer and transportation organizations in place, production was prepared as a Heinkel family affair. Relatives operated small factories in Swabia and were eager to produce and supply scooter components to the patriarch of the family. Sales started with 25 units per day, with only one model and

109

Fig. 6.1. Heinkel touring scooter (1952).

one color, green (some people called it giftgrün = poison green). The name was
touring scooter "Heinkel Tourist," the emblem the famous one of Heinkel's aircraft,
a capital "H" floating on an ocean wave and adorned with a four-prong wing. To
eliminate bogus parts in case of warranty claims, the trademark was cast and/or
inscribed on all proprietary engine and scooter parts.

It was a great day for Heinkel and his associates when the first scooters rolled off
the assembly line. The scooter had to be bought at a dealer but could be picked up
at the factory. The direct contact with customers was enjoyable as well as
harrowing. One day, a police officer of a neighboring town familiarized himself
with the scooter he had bought. He drove it inside and outside the plant, on the
autobahn and on field roads. He really liked the vehicle. "You see," he told me, "I
am so finicky because I'll use the scooter on my vacation." I said, "That's a great
idea. Where will you be heading?" He replied, "I'll drive through France and Spain
to Gibraltar, cross over to Morocco, drive along the Mediterranean coast to Tunis,
then relax on the boat to Italy, drive along its boot and return home via Switzerland."
Keeping a straight face but shuddering inside, I mulled over my options for this
situation. They were: (a) tell the man he takes an unusually high risk with this trip.
In case something happens, we cannot as yet provide service along this route.

(b) Say nothing and take the rap in case he gets stranded in the North African desert, surrounded by hostile Berber tribes, blaming me for not having warned him. Finally, (c) trust our scooter, wish him good luck — and us as well. I chose the last option. Lo and behold, after four weeks, the officer drove into our yard, smiling. He said it was a great trip, no problem whatsoever. What a relief to all concerned!

Another less serious incident: A young attractive lady refused to accept her scooter with an obvious fault. We were, however, unable to find the defect. She then pointed to a small blister of glue on the name "Tourist." It apparently had squeezed out when the plastic sign was attached to the front body panel. After surgical removal of the blob, she was satisfied and drove away with a grin on her face.

Our Heinkel Tourist soon made headlines in the newspapers and the trade press. Male and female drivers won prizes at national and international rallies from the Austrian Alps of Carinthia to the flat roads of the Isle of Man. It became as popular as Heinkel had predicted. Consequently, production rate could gradually be increased.

To enhance the touring aspect as well as the convenience and comfort appeal, we successively introduced advanced features: a more powerful 175 cc engine with 9.2 hp at 5500 rpm, electric starting, a four-speed gearbox and larger wheels, 10-inch rims instead of the original 8, resulting in a top speed of 60 mph and an only slightly lower fuel economy of 102 mpg. Empty weight increased to 320 pounds from 230. Even in those good old days, we had to provide a theft-prevention lock! Beautification included a variety of color schemes and aluminum or chrome trimmings. In addition, the dealers offered a host of extras, from plastic windshields to sidecars. The starter was combined with the generator in the engine flywheel. The location of two 12-volt batteries instead of one 6-volt posed some design problems. Finally, we put them in the lower portion of the luggage compartment, where the wires to the starter were very short and inadvertent acid leakage would not harm the driver. A patent was granted for this ideal arrangement.

The electric starter installation led to a hilarious occurrence. The head of Marketing phoned me with the complaint: "Your new scooter starts by itself." I replied, "It's not April Fools Day, cut it out!" But he was serious and offered a demonstration outside his office. A scooter was parked there on his stand, unattended. After a few minutes, the engine started to slowly turn over and then run idling, normally. Unbelievable if you did not see it yourself. Fortunately, there was a simple explanation. The small box near the steering column for the ignition/starter key was not watertight, rain had seeped in and short-circuited the starter contacts. When we put the supplier through the same routine, he was as surprised about this self-starting feature as we were. He gladly added a rubber seal to his box and retrofitted the scooters already in the field.

With all these improvements, Heinkel Tourist became the King of the Road despite its price tag of 1850 Marks ($440) and higher if the customer opted for all extras. Daily production rose to close to 100 units.

Shifting gears required an easily acquired skill but it was a nuisance for many, mainly female drivers. We therefore investigated automatic transmissions. The American automatics for passenger cars were out of the question for cost and fuel economy reasons. Inventors offered continually variable transmissions (CVT). We studied a few of them, two in particular.

The Beier transmission consists of two rows of intermeshing thin disks, one on the input, one on the parallel output shaft. One set of disks is tapered, the other has a T-section with the opposite taper. The change of the center distance of the shafts varies the transmission ratio. Torque is transmitted through the thin oil film of the oval contact areas of the disks. I remembered an article in an engineering magazine in which Dr. Beier had illustrated the adherence of the oil molecules to the metal surfaces whereas differently oriented molecules provided the shear strength for the transmittal of the force from one disk to the other.

A company in Karlsruhe manufactured Beier transmissions for industrial applications particularly for dough-making machines[6] and retained Dr. Beier as a consultant. When one of my design engineers and I paid a visit there, Beier handed me the key to a Volkswagen which was equipped with an automotive version of his transmission and invited us to drive it, which we did. On our return, he awaited our judgment with great anticipation. He was obviously disappointed when I simply said, "It moved!" At that time, we noticed a layer of disks on a reference table. The Chief Engineer explained, "These are all failed disks. For about 95 percent we have determined the failure cause, lack of lubrication, unsuitable fluid and/or impurities in it, insufficient axial pressure on the disks, etc. The remaining disks elude our scrutiny, we have no explanation whatsoever." Our trip to Karlsruhe had thus been worthwhile. This transmission type was of no use to us.[7]

Hydrostatic axial piston pump/motor transmissions were employed in many applications. Dr. Heinrich Ebert of Fürth, Bavaria, had introduced special patented

---

[6] A version of a Beier-type tapered-disk transmission had been employed in the Napier Nomad aircraft compounded diesel engine.

[7] Transmission of a torque through a thin oil film appears to be a high-risk design concept. Toroidal-drive CVTs comprise toroids on the input and the output shafts which are connected with a set of rollers. The contact radii of the rollers with the trochoids are varied for a continuously variable transmission ratio. Again, thin oil films provide transmission of the mechanical forces. They work reasonably well in constant-speed drives but not in vehicle applications. The most famous example was the Hayes transmission in Austin cars. It suffered from a sporadic service record and was discontinued. Improved versions for road vehicles were vigorously pursued in the early 1960s in the UK and the U.S. None of them achieved mass production.

features which made them suitable for road vehicles. One was installed in an NSU scooter, one of our strongest competitors. Ebert took us for a comparative test run on hilly and winding roads, the NSU with his transmission and a Heinkel Tourist with a three-speed gearbox. Despite our superior engine power, we could hardly keep up with the NSU/Ebert scooter. I cannot recall ever having driven a better automatic transmission; it had all desirable features, smooth, responsive, amenable to override for acceleration and/or braking — just perfect. Impressed with the Ebert drive, we conducted a preliminary design study for our scooter. Simultaneously, NSU had put a small series on the market. Unfortunately, there was a drawback — cost. This lead NSU to discontinue the program and so did we.[8]

In the wake of the scooter success, Heinkel wished to enter the low end of the two-wheeler market, first with an auxiliary engine (Hilfsmotor) to be attached to the rear hub of a bicycle, then a Moped itself.

In view of the low speed of a bicycle, I designed for the Heinkel Hilfsmotor an air-cooled cylinder with short, interrupted, offset cooling fins. They served their purpose well and, in addition, provided a distinctive picture compared to the other orthodox cylinders (see Figure 6.2).

Inventors often approached Heinkel with their ideas; he was famous for listening to them and sometimes helping them. One presented a design of a Moped with two advantageous features: a hollow frame including the gasoline tank, cast in aluminum, and an engine cum two-speed gearbox that could be operated with the flip of a finger. We subjected the frame to rigorous fracture testing, found it okay and accepted it for production. There we encountered one serious shortcoming: only one foundry would be willing and able to cast its thin walls. Thicker walls were not permissible for weight reasons. The engine and gearbox caused many operational problems; we were obliged to design our own. It was a challenge: a Moped two-speed gearbox has to fulfill an astounding number of requirements. We diligently pursued all avenues and finally arrived at a gear/clutch arrangement that offered the greatest convenience to the driver.

To complement the sturdy, heavy touring scooter, Heinkel asked for a simpler, lighter version. Emphasis was less on performance and more on styling. With the adage in mind, "Beauty lies in the eye of the beholder," I was reluctant to establish the final shape of the sheet-metal body myself. So I asked Heinkel to generate an attractive contour on a clay model with a hunting knife, and he did so with gusto.

---

[8] Ebert tried his hydrostatic drive for less cost-sensitive applications, e.g., a heavy truck at Daimler-Benz and an aircraft auxiliary drive at the Stratos Division of Fairchild on Long Island. The truck performed well. A new obstacle occurred — noise. High fluid pressure was necessary to keep size and weight within acceptable limits and the discontinuities of the fluid flow generated an intolerably high noise level.

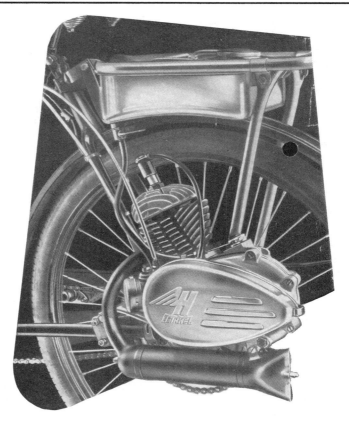

Fig. 6.2. Heinkel Hilfsmotor.

After some shaving of the clay, and looking, and shaving, and looking, he liked the shape and said, "That's pretty good now, isn't it? Go ahead and order the stamping die."

We proudly presented the first prototype scooter, the body in brilliant color, to Heinkel to admire. After one look he said, "Which idiot came up with this ugly shape?" On the tip of my tongue was the short answer, "It was you," but I chose a milder tone and reminded him diplomatically of his styling efforts with the hunting knife. After a few further looks at the scooter, he stated, "It's quite a pretty sight, isn't it?" We all nodded our agreement and were glad that the scene ended on such an amicable note. The customers agreed too; the scooter did fairly well.

All two-wheelers are a pleasure to drive in fine summer weather, but they provide little protection in the inclement seasons. That obvious fact gave rise to the *Kabinenroller* (cabin scooter). Messerschmitt, Heinkel's formidable competitor in the aviation field, had designed and built the first "tandem cabin" on two wheels with a retractable small wheel on each side. Others followed with side-by-side seating, two wide-track wheels in front and two narrow-track driving wheels on one shaft in the rear, and a front door. Italian and German models, some built under license, were popular.

Heinkel wanted to be in this market, too. Realizing that time was of the essence he proposed to use the Tourist powertrain and build a cabin around it. I was against this expediency, expressing my concern about the vehicle's stability, particularly in an emergency. Such a three-wheeler was, in my humble opinion, too dangerous for most people to drive. Also, small four-wheeled cars already were replacing cabin scooters at an increasing rate. His airframe people prevailed and designed the cabin. The powertrain required some modifications, which I provided reluctantly.

The Heinkel cabin scooter *Kabine* was unable to match the success story of the Heinkel two-wheeled scooters. My prediction became true, production in Zuffenhausen had to be discontinued, and that under license in Ireland faded out, too.

Developing, manufacturing and marketing of engines and vehicles for the middle class of consumers was a drastic change from the sedate, almost leisurely pace of the government-sponsored tank engine development in England. It was not smooth sailing by any means; we had our ups and downs. We were subjected to continuous pressure from two sources: first, from the market conditions generated by the buying public and the competition, and second, from Heinkel himself, including outbursts if our progress was unsatisfactory to him, or late, or both. Joyous exhilarating events alternated with frustrations. In my opinion, we had "zu viel Heu auf einer Gabel" (too much dirt on one shovel). The experience in the competitive, commercial field was, however, worthwhile. It offered a wider view of the engineering profession and an appreciation of the challenge every field presents in different ways.

The commercial consumer industry depends on paying customers; there are essentially two determining factors: the maker and the user. In contrast, the military and space industry contains three factors: the maker, the user, and the government, which pays both the maker and the user. The payer is represented by the legislative branch of the government and finally the taxpayer. In this three-sided arrangement, the maker and the user work closely together, which somewhat limits the influence of the payer on the management of the projects. That fact prompted President Dwight D. Eisenhower to warn his nation in his farewell address about the power of the military-industrial complex, the maker and the user. In the commercial market, user and payer are the same, and the market forces provide the balance between the maker and the user.

The moped and scooter market declined to almost zero in the late-1950s and early-60s. New requirements on fuel economy and exhaust emissions began to revive it in the 1980s. I looked with amazement at the colorful brochures of scooters and cabin scooters coming from Japan and European countries. They resemble in many ways what we had designed, developed and marketed in the mid-1950s.

Gas Turbines, Aviation and Space

Heinkel savored his commercial ventures. His heart, however, was really in aviation which he had vigorously pursued since he had flown and crashed his self-built aeroplane in 1911 in Stuttgart. Like a magnet he had now attracted to Stuttgart a microcosm of German aviation, from aerodynamicists, designers of aircraft engines and frames, machinery and tools, general and patent attorneys, to men of all trades. In addition, Heinkel had personal representatives in many industrialized and third-world countries who rendered their services to him. With this international background, he tried to market or sell licenses of in-house and other inventions.

He often quoted with great personal pride his success in selling licenses for the *Sprengniete* (explosive rivet), an invention of two of his engineers, the brothers Karl and Otto Butter. This rivet permits joining of sheet-metal panels where one side is inaccessible. On that inner side, the rivet contains in a recess a small amount of gunpowder which, by heating the outside rivet head, explodes and forms the inner rivet head. The explosive rivets had been fully developed in cooperation with the Dynamit-Nobel-Konzern. The German air ministry failed to see the rivet's potential and allowed Heinkel to license it to foreign countries. Japan had taken the first license for 1.2 million Marks. The sale to the American Du Pont Company had become a dramatic affair. It was conducted in early 1940 during the *Sitzkrieg* (non-shooting war on the Western Front) by telephone from Milano in neutral Italy to New York City. On the phone in Milano were the American consul, a director of a Milano bank, and Heinkel's representative with a package of the relevant licensing documents. In New York were the German consul, a director of the Deutsche Bank, and a representative of Du Pont with a certified check for $250,000. After introducing the individual participants and inspecting the documents, first the check was handed over in New York, then the document package in Milano, and the deal was thus completed by phone.

Other examples were: a polishing machine for gas turbine blades; for multifuel combustion a fuel injector with a finned chamber for heating and evaporating fuel prior to its injection; a rivetting machine for airframe panels; and conventional and novel gas turbines for aircraft propulsion. I had to spend — and sometimes waste — a lot of time assessing their viability and prospects for a user or a partner for their development.

The rocket development and application during WW II had spawned the peaceful exploration of space. Heinkel wanted to participate in these ventures. After all, his planes were first to fly with rocket propulsion (first with rocket-assist and then with the pure rocket-propelled He 176). As a token, he donated the Ernst Heinkel Award for the best paper of a space conference held in May 1952 at Stuttgart. I was asked to attend the sessions. I found somewhat distressing a paper investigating the number of rocket stages to achieve earth orbit. Its author treated the subject with lengthy mathematical formulae paying, in my opinion, only scant attention to the different design features of the various stages. Reporting to Heinkel on the conference and the lack of realistic thinking, I quoted this paper as an example. Heinkel was unable (or unwilling?) to attend the award ceremony and asked me to represent him. I agreed, reluctantly. The final irony was that I had to present the Ernst Heinkel Award to the author whose paper I liked the least. This incident may have subconsciously contributed to my shying away from space engineering during my career.

Max Adolf Müller was again on the scene with a gas turbine employing thermal compression, an idea originally conceived by Hans von Ohain. This concept had been theoretically and experimentally investigated as a special project at Heinkel. Other investigators had done similar work and published it. Thermal compression eliminates or reduces the usual compression with turbomachinery. Component tests had confirmed the principle but also revealed the difficulties in achieving the expected performance in practice. Knowing I was familiar with von Ohain's experiments, Heinkel asked me for my consulting services. They were just that; Max Adolf Müller carried full responsibility for the project.

A personal invitation of the Society of British Aircraft Constructors to their 1952 Farnborough Flying Display and Exhibition pleased Heinkel immensely; he gladly accepted it. Familiar with the English scene, I was to accompany him on this historic journey. Despite his outstanding achievements in aviation, he hated flying,[9] so we did the trip by car, a Mercedes 220 into which five people were squeezed: Dr. and Frau Heinkel, their 13-year-old son Karl Ernst, Herr Lahmer (the chauffeur), and me. We arrived via Heidelberg and a night's stay at the Quellenhof in Aachen, and arrived the next morning at Ostende. Our car was hoisted into the ship's cargo hold. We enjoyed crossing the English Channel. An annoying incident somewhat spoiled our arrival on British soil in Dover. The Mercedes would not start, even after many attempts and Lahmer's experience. Bystanders who had noticed our German license plate, grinningly enjoyed the, for us, frustrating scene. A ship's helper came

---

[9] He never forgave the British authorities for having him flown to London after the war for interrogation.

to the rescue advising Lahmer to open the shutoff valve in the gasoline line. For safety reasons, the ship's crew had closed it. Lahmer never used it and had completely forgotten its existence. When it was opened, the car started normally.[10]

Reminding Lahmer at regular intervals about driving on the left side, we arrived safely on Sunday afternoon at the Park Lane Hotel, Piccadilly, London, W 1. A former associate of Heinkel, A.R. Weyl of the Dart Aircraft Company, welcomed us there. Heinkel thanked him for his arrangements and presented a few gifts: Swabian Wurst, Schinken, Bauernbrot and Wein.

Our arrival had been previously announced with a newspaper headline, "Luft-waffe's 'Big 3' will see Britain's latest types," referring to Messerschmitt, Heinkel and Dornier. No wonder, a *Daily Express* reporter interviewed Heinkel shortly after our arrival. Under next morning's headline, "Heinkel in London - at zero feet" and "Build Comets in Germany? Oh, jawohl,"[11] appeared, after a short introduction, a vivid account of the encounter as follows: Professor Ernst Heinkel had left his visiting card on the city before, dropped by hundreds of Heinkel bombers during the last war; this time he was more amiable. The Professor has something to offer: Heinkel could build the planes Britain and Europe needs. Heinkel was ahead of the world on jets. He had a jet flying in 1939, nearly two years before Whittle's, and a twin jet in 1940.

On Monday, we enjoyed some sightseeing — Parliament, Westminster Abbey, Trafalgar Square, St. Paul's Cathedral, and the Imperial War Museum, where Heinkel saw for the first time a V2 rocket. I also showed the Heinkel family the small bombed section of the city as well as the Piccadilly Circle subway station where the people had taken refuge during the London Blitz without being bothered by his He 111 bombers.

The next day, SBAC officials welcomed Heinkel and his entourage kindly and amicably. We encountered the same pleasant atmosphere at the hospitality chalets. There I took the opportunity to introduce Heinkel to industry executives I had worked with after the war. He enjoyed the static and flying display, the Comet in airline livery, the AVRO delta-wing Vulcan bomber, the Armstrong-Siddeley twin-jet Canberra and De Havilland DH 110 fighter. He and his son snapped many

---

[10] A most embarrassing experience of a Mercedes not starting occurred during the visit of Queen Elizabeth II to Stuttgart. From Stuttgart's TV tower, Her Majesty enjoyed a breathtaking view of the Swabian capital, including the Daimler-Benz headquarters and factory, and then took her seat in the car, which could not be made to start. The case was not as simple as ours on the Dover pier. The Queen's Mercedes was equipped with electrically operated windows, something new to the British chauffeurs. They had played with this gadget long enough to drain the battery so that it was unable to start the engine.

[11] The De Havilland Comet had made the world's first commercial flight of a jet airliner on May 2, 1952. The plane was ordered by many airlines and, thus, was a major attraction of the SBAC show.

pictures, and so did reporters who also questioned him. Subsequent newspaper articles quoted him as follows, "I am delighted with everything," "Britain's planes - best in the world," "I'll make the Comet. You don't build fast enough," "There is a lack of energy and production capacity." The latter concern was expressed in a general headline, "There's plenty in the window - but so little under the counter."[12]

After a few days, Heinkel got restless and wanted to go home. I told him that could be easily arranged except for the car and his adversity to flying. I was right, the ferries were booked solid for weeks to come. He persisted with the remark that I will be smart enough to find a solution. Recalling the Heinkel newspaper headlines and pictures gave me an idea that might work — and it did. Asking him to let me handle everything without his interference, he and I went to the Royal Automobile Club headquarters. I asked for the General Manager. On his arrival I said to him, "Sir, may I introduce Professor Dr. Heinkel." The gentleman was overwhelmed by the unique opportunity to shake hands and chat with such a famous visitor. After some polite conversation, he asked the question I had been waiting for, "Can we be of service?" Shortly afterwards, Lahmer had Heinkel and his family in the car heading for Stuttgart. At 3:30 a.m. the next day, the telephone in my hotel room awoke me, "Here's Heinkel. We're home safely." He thanked me for a job well done.

When I returned to Frankfurt by air, my wife and Lahmer greeted me with the question, "Did you see it?" They had heard on the radio the tragedy of the De Havilland DH 110. The plane had disintegrated after a supersonic dive and plunged into the spectator crowd. A sad ending to a great show.

Heinkel reported on the Stuttgart radio about the impressions of his Farnborough journey and offered some suggestions. It is nonsense for five Western States to design ten different types (of aircraft). We should have as few types as possible. This goes for commercial as well as military aircraft, he said.[13]

This interview and subsequent coverage in the press put Heinkel in the forefront of a revival of the German aircraft industry. This type of work was still verboten but lifting of this restriction appeared imminent. With this anticipation, Heinkel started to hire jet propulsion engineers who had left Germany after WW II and tentatively began design and development of a powerful turbojet. It was a high-risk project due to the limited or nonexistent availability of testing and manufacturing facilities.

---

[12] I kept smiling, inwardly. After WW II, Heinkel had lost all airplane manufacturing facilities and no plans for new ones existed.

[13] Heinkel was ahead of the times. It took almost twenty years before the UK, West Germany and Italy put his suggestions into practice with the joint development and manufacture of the multi-role combat aircraft (MRCA) "Tornado." The four-nation European consortium Airbus Industries represents a commercial example.

Germany's automotive industry executives also were interested in the gas turbine developments. One of these was Dr. Ernst Mahle (1896-1983), the technical director of the renowned Mahle piston factory which had been founded by his brother Hermann Mahle (1894-1971) and Hellmuth Hirth after World War I. Since my days at Hirth-Motoren, our relationships in the technical and personal field made for a good chemistry between Ernst Mahle and myself. Knowing me pretty well, he asked me for advice on solving a dilemma he had encountered in the recent past, i.e., to build up his piston business or to start with the manufacture of gas turbine blades.

The Mahle manufacturing facilities had just recovered from the ravages of the war and he had to invest heavily in modern machinery. He had followed the progress of the jet engine, the proud announcements of automotive gas turbines, the actual demonstrations of the Rover gas turbine-powered automobile and the Kenworth truck driven by a Boeing gas turbine. All newspapers and trade magazines were full of praise for this prime mover of the future. His customers, too, expressed concern about the crossroads of piston and gas turbine engines.

He was familiar with my jet engine work at Heinkel-Hirth during the war and assumed I had been engaged in gas turbine development in England. With his and my background in mind, he asked me for a briefing on gas turbines in general and automotive applications in particular. I thanked him for his trust in my expertise and in my judgment on the potential and future of the gas turbine. In my naivety, I expected a face-to-face give-and-take discussion with him alone, and accepted his invitation to see him before Christmas (1952).

After greeting me cordially in his office, he said, "Let's go," and opened the door to his conference room. There, his staff — Research and Development, Manufacturing, Marketing, Sales and Finance — applauded my introduction, and were eager to grasp every word from my lips. Of course, I had neither prepared notes nor slides and made the best of it under the circumstances.

Starting with the fundamentals of the thermodynamic cycle, I explained its sensitivity to the various parameters, particularly stressing the effects of size and power level, as well as the physical and technical difficulties in materials and manufacturing. The ensuing specific characteristics in the high-, intermediate- and low-power class led to some individual applications without showing a distinct pattern yet. Then I turned to the main interest of the Mahle enterprise, the vehicular gas turbine.

In contrast to other successful applications of the gas turbine, its operation in a vehicle poses additional formidable difficulties. Also, the competition of the well-established Otto and Diesel engine can hardly be overestimated. Fuel economy over the full speed and power ranges is a paramount objective. This may be accomplished with high cycle temperatures in combination with high pressure ratios *or* with heat exchange from the exhaust to the compressed air. The latter method appears more promising. Even then the gas temperatures are too high for materials used in piston engines. Either high-temperature, expensive alloys *or* air-cooling has to be employed. Cooling the turbine rotor detrimentally affects specific power and fuel consumption to an increasing degree with decreasing turbine size and part-load power. Already during the war, efforts were in progress to replace the strategic alloys and air-cooling with ceramic and metal-ceramic materials.

All of these considerations are reflected in the recent history of the gas turbine. It is predominant in the military and commercial aviation (though not in general aviation). Applications emerge in the 1000- to 5000-horsepower class, and enormous difficulties arise in the low-power class and particularly in the automotive field, notwithstanding the glamorous demonstrations. I concluded my talk with the statement, which I put in writing later on, "Besides, since all attempts to reduce the fuel consumption will result in increased weight and volume, abandonment of the mechanical simplicity and facility of inspection, as well as increased susceptibility and shorter overhaul intervals, it is in any case doubtful whether the gas turbine will be able to replace the piston engine in the low-power range. There are few characteristics of the gas turbine in this power class which cannot also be achieved by piston engines and auxiliaries; the gas turbine configuration leads to many problems in the manufacture, the materials, operation and installation, the solutions of which cannot yet be estimated. It will, therefore, be useful to follow the development of gas turbines in the lower range of output very carefully."

Ernst Mahle was well pleased with my talk, obviously with a sigh of relief, and rewarded me with a generous honorarium. He subsequently asked me to put my speech in writing for publication in the *Mahle Journal*. Appendix 6.1 presents his Introduction in German, and 6.2 the full treatise translated in English.

Noteworthy is a comparison of the above assessment of the automotive gas turbine with the one I had made five years earlier. In simple terms, it reflects theory vs. practice. The details of the aero/thermodynamic layout and design, as well as the experiments, had brought to light all forms of losses: aerodynamic, gas leakage, heat, parasitic, and what have you. These still plague automotive gas turbine engines.

My 1952-53 Mahle paper, including the forecast for the automotive gas turbine, proved on-the-mark for a long time — and still does. The Mahle Group has since produced millions and millions of pistons and not a single gas turbine blade. I occasionally use this short article as a base for papers and presentations on the subject.

Heinkel's Farnborough publicity also prompted him to finish and publish his memoirs. He was dissatisfied with the editor he had engaged but had found an excellent replacement. The 561-page illustrated book was published in the Fall of 1953 with the title page: Ernst Heinkel - Stürmisches Leben - Herausgegeben von Jürgen Thorwald (Stormy Life - Edited by Jürgen Thorwald; an English translation was published bearing the title "He 1000"). Heinkel had been warned to be careful and not to annoy or embarrass anybody, otherwise he may be in the same shoes as Anthony Fokker (1890-1939) had been. Within a week of the publication of his memoirs, Fokker was faced with half a dozen lawsuits. Heinkel was spared such misfortunes. However, he displayed his frustration in having been denied mass production of jet engines. Not a single name of the Heinkel-Hirth team is mentioned — including mine.

Heinkel Jet Engine He S 053

Heinkel's most ambitious post-war project was the design of a jet fighter and its engine. The engine was designated He S 053, indicating that it was conceived in the year 1953. With a thrust of 6500 kilopond (14,330 lb or 63.8 kN), it was extremely large for the time, but in its layout fairly conventional. It borrowed features from Heinkel's He S 011 and SNECMA's ATAR engines. The engine consisted of an all-axial eleven-stage compressor with variable inlet guide vanes, a combustion chamber with nine cans and a following annulus, an air-cooled two-stage turbine, and a two-position jet nozzle. A specific fuel consumption at sea level of 0.93 lb/lb-hr was expected. An afterburner later to be incorporated would have increased the thrust to 9000 kilopond. Due to the scant availability of test facilities, the engine development progressed slowly and, to the best of my knowledge, a complete engine test was never conducted. The fighter design did not reach the hardware stage. Finally, the whole project was terminated by the original sponsor.

To demonstrate his dedication to modern aviation, Heinkel invited the old and new pioneers in this field to a gathering in his just-finished company Kasino, and they all came. He had the great pleasure to welcome among the 40 scientists and engineers Henrich Focke (1890-1979, helicopter), Georg Madelung (1889-1972, aircraft design, catapult experiments and operation), Eugen (1905-1964) and Irene (1911-1983) Sänger (rockets and space exploration),[14] Arthur Weise (1904-1978, flow dynamics), and others. Madelung talked about the American superiority and the lack of sufficient funding to catch up with the U.S.

The star of the illustrious meeting was Siegfried Günter, who had recently returned from Kubishew in Soviet Russia. Everybody agreed that the Russian fighters of the MIG series showed his influence and mark. He just smiled. He would

---

[14] In the late 1980s, the advanced European two-stage space transport system was named *Sänger Projekt* in their honor.

not publicly admit that, probably being cautious for fear of possible Russian denunciations. Respecting his right to privacy, nobody pressed him further on the subject.

I was privileged to present a treatise on the past, present and future of jet engines, and their characteristics.

Curtiss-Wright Overture

To expand his effort in the aircraft engine business, Heinkel had succeeded in interesting Curtiss-Wright in joint ventures. Roy T. Hurley (1896-1970), then Chairman of the Board and President, wanted to expand its jet engine R&D and extend the life of its successful reciprocating engines with various programs. Heinkel's negotiations on a novel jet engine concept led to no tangible results and were finally abandoned.

For the dieselization of its radial, single-row Wright Cyclone aircraft engine, Curtiss-Wright looked in the mid-1950s for help in the Stuttgart area. The Bosch, L'Orange and Daimler-Benz companies had acquired a lot of experience with diesel engines and their injection equipment; I enjoyed a good relationship with them. Hurley and his entourage conducted discussions at these companies, and then with us at Heinkel.

I particularly enjoyed meeting the engineers again who had interrogated me in 1945. We reminisced about these times. They also became interested in my doings since then and finally asked me to let them have my resume. They might be able to offer me greater opportunity and put me to better use than Heinkel, they said. After about a year, their European representative visited me in Stuttgart and first asked why they had not received my resume. I answered, "It's very simple, I never sent it." He said they were very much interested in my services now. Curtiss-Wright suggested inviting me for a visit to the United States for a first-hand look. I did that in June 1955. Curtiss-Wright's size and performance impressed me, with sales of some 500 million dollars and close to 30,000 employees in 17 divisions that all worked on air transportation and in associated fields. Most of the time I spent at the Wright Aeronautical Division (WAD) which developed and produced reciprocating and jet engines as well as ramjets. The engineering organization pleased me, with Lundquist as Director of Engineering and Farrar as Chief Engineer. My position would be Staff Engineer in the Design Department. There was one drawback in our discussions. Since I had no security clearance, the advanced projects, twin-spool jet engines, a dual-cycle engine comprising a jet engine and a ramjet, and liquid rocket engines could not be shown to me. Thus, I was unable to fully assess the relative position of Curtiss-Wright's engine technology to its competitors in the U.S. and abroad.

Home again, my wife and I mulled over my options. In Germany the jet engine business would be for various reasons restricted to production under license, with its own developments far in the future. In short, it would be extremely difficult to put my knowledge and ability to fruitful use. With the help of the saying, "The past is prologue," we decided to accept the Curtiss-Wright position I had been offered during my visit.

Heinkel congratulated me on this step, admitting that I would find in the U.S. conditions more favorable for my capabilities than he was able to offer. He regretted having to let me go, thanked me for my loyal service with a toast (Figure 6.3), and wished me and my family the best for the future.

Immigration to the U.S. involved, at that time, complex procedures on both sides of the Atlantic. Curtiss-Wright had to prove that an American engineer equivalent to my experience was unavailable. Military clearance also was required. The scrutiny of my family and myself came close to that of a presidential appointee to

Fig. 6.3. Heinkel (left) toasting me before my departure for the U.S.

be confirmed by the U.S. Senate. After interviews and physical examinations at the American Consulate General in Munich, we finally received our immigrant visas. At farewell visits and parties we were, in addition to good wishes, showered with presents. Two were especially appropriate, a 2x3-foot map of "The Circle of Swabia, dedicated to the most Serene and Sacred Majesty James II" (1633-1701), and a 1-3/4 x 2-foot map *Territorii Vlmensis, Nova et Accurata*.

At the Stuttgart Central Station, colleagues, fraternity brothers and sisters, friends and relatives saw us off for the night train to Paris. After a day of sightseeing we took the boat train to Le Havre where we embarked on the SS United States, the world's fastest ocean liner and at that time the sole passenger boat with full air-conditioning. The life-saving drill the next morning was conducted with military precision and discipline. Members of the band bitterly complained that they had to go through this routine on each and every trip. They probably regretted their grievance when a few days later the Swedish liner, Stockholm, rammed the Italian Andrea Doria, which sank and lost 50 lives, some of them due to inadequately prepared and conducted lifesaving procedures. In the early morning of July 17, 1956, we admired the south shore of Long Island, the Statue of Liberty, and were welcomed in New York City by a CW official.

Until we found a home in Ridgewood, New Jersey, we stayed in a hotel in Manhattan. I commuted to the Wood-Ridge plant with a company car. We savored the opportunity to adjust to the American way of life, to visit the historic sites and monuments, and to walk in Central Park on cool nights, sometimes barefoot.

Appendix 6.1

MAHLE

M A H L E   K O M M . - G E S.   S T U T T G A R T - B A D   C A N N S T A T T

Pragstraße 26-46 - Fernruf 502 46 / 528 41

Die Gasturbine, heute und morgen

Die Frage, ob die Gasturbine auch für das Kraftfahrzeug von Bedeutung ist, wird seit einigen Jahren immer wieder diskutiert. Da dieses Gebiet noch neu ist, haben wir einen ausgesprochenen Gasturbinen-Fachmann gebeten, uns in einem Aufsatz seine Meinung darüber einmal aufzuschreiben.

Wir glauben, daß dieser Aufsatz aus berufener Feder viele Fachgenossen interessiert und daß es Ihnen eine Freude bereitet, wenn wir Ihnen diesen Aufsatz hiermit überreichen.

Ernst Mahle

Stuttgart-Bad Cannstatt, im März 1953

127

Appendix 6.2

Translation of "Die Gasturbine, heute & morgen."

The Gas Turbine, Today and Tomorrow

The question whether the gas turbine is also of importance for vehicular applications has been discussed again and again during the last several years. Since this area is still new, we have asked a renowned gas turbine expert to express his opinions in an article.

We believe that this treatise by an expert will be of interest to many colleagues in this field and that you will appreciate receiving this article.

Stuttgart-Bad Cannstatt, March 1953   Ernst Mahle

Toward the end of the war and after the war, interest in gas turbines became predominant because of their application in aviation. The success thereby gained prompted various parties to hope and wish that the gas turbine would quickly and overwhelmingly succeed in other areas, too, and appropriate development work was begun at many locations. Abroad, great expectations were based on turbines which were fully operational and competitive with other engines, and it was probably for this reason that Germany was prohibited from participation in this area of technology.

Developments during the last few years have not confirmed the early highly optimistic hopes, at least not in the originally intended areas of application, and it is interesting and of importance for estimating future trends to investigate the reasons. In this regard, it is appropriate to differentiate between those problems governed by the laws of physics and the technological and economical ones.

Operating Process

The vast majority of the gas turbines recently developed and planned operate with the constant pressure method (Brayton cycle) and consist of compressor, combustion chamber, and turbine sections which may be extended by a heat exchanger for the purpose of utilizing the hot exhaust gas energy. The mechanical energy produced in the turbine is partially extracted for driving the compressor, the remainder is available as useful energy. Output and compressor energy are currently of the same order of magnitude. Therefore, the process is also described as a "differential" process and the compressor work as negative work. The most important characteristics of the thermal engine, i.e., the specific output (effective output per unit airflow) and the thermal efficiency, are dependent on the turbine inlet temperature and increase with it. However, because of this energy distribution, the qualities become very sensitive to the losses which are associated with the individual processes, i.e., the compression, combustion, expansion, and heat exchange; and the influence is naturally greater if the energy processed by the various sections is higher. For turbomachines, these losses are usually expressed by efficiency, for the combustion chambers and heat exchangers by pressure, heat, and leakage losses.

As is well known, these circumstances postponed the successful development of constant pressure gas turbines until such time as the results of the aerodynamic, thermodynamic, and metallurgical research had decreased the losses to acceptable values and allowed the use of sufficiently high gas temperatures.

Turbomachines have two sources of losses: the purely aerodynamic losses and the losses caused by gaps, surface roughness, and flow short-circuits; the first source is considered to be revaluable, the second source is not. The revaluable losses are dependent on the Reynolds number under otherwise identical conditions and increase as the size of the engine decreases. The losses that are not to be revalued also increase as the engine size decreases since it is difficult and relatively expensive to retain the same quality of blade profiles, surfaces, gaps and clearances, and mechanical quality. For these reasons, thermal efficiency and specific output decrease as a turbine engine is reduced in size and finally reach such low values that it is impossible to operate it economically. At the present time, this low limit lies in the range of 100 hp. Gas turbines for smaller output of approximately 50 hp are being built for special purposes, however, they exhibit an extremely high fuel consumption (on the order of 2 lb/hp/hr) at full load.

The losses in the combustion chamber and heat exchanger sections are mainly pressure losses caused by gas friction, deflection and flow separation, which are subject to the mean flow velocity and increase with it. Since the mean flow velocity determines the geometric flow area, the losses are dependent on the allowable engine volume (bulk) and weight and, therefore, also on the acceptable cost. These losses not only influence the thermal efficiency and specific output, but are also decisive for the starting process and the idle operation of the engine.

Required output and available engine volume therefore determine extensively the specific characteristics of the gas turbine, whereby the contributing physical influences can be reduced only insignificantly by engineering measures.

Fuels

Constant pressure (Brayton cycle) gas turbines operate with continuous combustion and are, therefore, inherently insensitive to fuel characteristics. Originally, this advantage over internal combustion piston engines was overrated since it was assumed that any gas turbine engine could use any type of fuel. However, in practice this caused difficulties, especially when using undistilled, high-boiling fuels, bunker and heating oil, etc., which caused deposits on the turbine blades, and corrosion. These difficulties imposed special limitations such as reduction of the maximum cycle temperature or application of high-temperature alloys despite the fact that these expensive and scarce materials were not required for reasons of strength and temperature.

On the other hand, considerable progress was made in the combustion of solid fuels such as coal dust, peat, sawdust, etc., i.e., with fuels which are not suitable for piston engines.

Combustion of gases such as natural gas and blast-furnace gas causes the fewest difficulties for gas turbines. The possibility to use cheap fuels is closely connected with the application and operating conditions and is reduced as the requirements for starting behavior, maintenance, overhaul intervals, and operating safety are increased.

High-Power Class

The physical conditions are especially favorable for the high-power class above approximately 2000 hp. It is, therefore, no mere coincidence that the first gas turbines built for commercial applications were in this class. For very high power, the problems of cost of the overall project, design and manufacture become predominant and set an upper economic limit which may be assumed to be in the range of several tens of thousand horsepower.

The specific characteristics of the gas turbine which become apparent in this power range, i.e., moderate manufacturing cost, low maintenance cost, minimal starting time, lack of vibration, and modest requirements in regard to fuel, are often sufficient to prefer it to the steam turbine, the steam engine, and the diesel engine despite its possibly higher fuel consumption. This is especially true for ship propulsion, locomotives, and for electric power generation.

In this regard, the construction can be more simplified if less emphasis is placed on low fuel consumption, e.g., for power generation during peak periods and for applications as a booster engine for acceleration of ships that are normally operated by diesel engines.

Intermediate-Power Class

For intermediate-power classes down to 1000 hp, there are also no fundamental physical and technical difficulties concerning the layout, design, and manufacture of gas turbines. However, the low fuel consumption of the Diesel engine, which dominates this power class, cannot be reached by the simple constant-pressure gas turbine. This type, therefore, has chances only in those areas of application where its outstanding characteristics as compared to the Diesel engine are of advantage. These are: low engine volume and weight; lack of reciprocating masses and parts subject to friction and wear, as well as lack of vibrations and oscillations; insensitivity to lubrication media; low maintenance requirements; and long overhaul intervals.

The advantages are especially attractive to rail cars, ships, portable power units, emergency power generators, pumps for water supply and for drainage, and prime movers in oil and natural gas fields.

Low-Power Class

As mentioned above, the relative losses generally increase with a decrease in power; therefore, a turbine in the power class of 100 hp will exhibit a considerably higher fuel consumption than a gas turbine of approximately 1000 hp designed and built under the same conditions. But even this already proverbially high fuel consumption of the small gas turbine may be accepted if its advantages as compared to the Otto and Diesel engines — namely the considerably lower weight and volume, freedom from vibration, good starting characteristics, and lack of cooling water — are decisive for its application. This may be the case for powering transportable ventilators, fire-fighting pumps, boats for patrol and police purposes, auxiliary and emergency power generators, etc. This type of gas turbine engine has also been used for road vehicle application. In this regard, the more favorable torque characteristic as compared to reciprocating engines is also a contributing factor. This is the result of separating the turbine section into one serving the compressor — the so-called compressor turbine — and the mechanically independent power turbine which drives the vehicle. Compressor, combustion chamber, and compressor turbine constitute the gas generator which furnishes the power turbine independently of its speed with effective gas energy resulting in a torque ratio between full speed and stall of 1:3. Unfortunately, this ratio is not sufficient to satisfy all requirements of road vehicles, and a small transmission gearbox is still required between power turbine and propeller shaft. For passenger cars, this can be a two-speed version; for trucks and tractors, several speeds would be required.

Gas turbines have been experimentally installed and tested in vehicles; initially almost simultaneously by Rover for passenger cars and by Boeing for trucks in the Spring of 1950. These vehicles represent movable test stands and are not yet competitive with the standard propulsion units. Other gas turbines intended for this type of application were already on test stands prior to these road tests.

The Vehicular Gas Turbine

The following will discuss in detail the mechanical and manufacturing problems peculiar to vehicular gas turbines which exist in addition to the problems already discussed. These considerations are also partially applicable to higher powers.

The rotational speeds of the gas generator and the power turbine increase as the output decreases and reach values of 60,000 and 40,000 rpm, respectively. These speeds constitute high requirements on the bearings and reduction gears between power turbine and drive shaft.

Because of the small size of the parts, it is difficult to produce the required smooth flow path and blade profiles. In the hot end of the engine, the high temperatures and temperature gradients also cause distortion of the parts which must be kept to a minimum in order to maintain the proper air and gas distribution in the combustion chamber and the small tip clearances between rotors and housings. For these reasons, the gas turbine must be considered as a unit, and repair of a gas turbine outside the manufacturer's plant or mere replacement of spare parts cannot be considered.

Materials

The turbine rotor is subject to high stresses at high temperatures which can only be accepted by high-temperature materials. Although the turbine nozzle vanes are not subject to centrifugal forces, they are subject to gas forces and vibrations as well as thermal changes, and they accept the full turbine inlet temperature and, therefore, also require appropriate materials. To a certain point, this also holds for internal parts of the combustion chamber for which the vibratory loads and temperature changes are decisive.

The development of high-temperature materials has progressed considerably and especially nickel-based alloys take an outstanding position here which has, however, lately been challenged by cobalt-based alloys. These alloys consist of high-value elements which are found and produced in relatively small quantities and which, as "strategic" materials, are not available in unlimited quantities. This scarcity of high-temperature materials might render the application of small gas turbines in large quantities questionable in their present configuration. It should, therefore, be investigated whether the requirements on the materials can be made less stringent and the need for alloy elements can be reduced. In Germany, these approaches were investigated at an early stage and good results were achieved by means of design and cooling techniques. In addition, starts were made in the use of ceramic and metal-ceramic materials and coatings. These directions in development were also pursued abroad and should be of importance to the small gas turbine. Closely connected with this question are the manufacturing methods and the cost of gas turbines which will be important factors for their possible introduction.

In this regard, the manufacture of turbine blades and their attachment in the disk may be the deciding task.

Fuel Consumption

The fuel consumption of the experimentally operational vehicular gas turbines is considerably higher than that of comparable Otto and Diesel engines, and is usually a multiple thereof. Therefore, attempts to reduce it have not been lacking. Primarily, there are two means available: The simultaneous increase of the turbine inlet temperature and compression pressure ratio or the installation of a heat exchanger. The first means requires the use of multiple-stage axial or radial compressors. The radial compressor is simpler but larger in diameter and probably has a lower efficiency than the axial compressor. The axial compressor is expensive due to the large number of exactly profiled blades, and is sensitive to dust, foreign objects, and tip clearance changes; however, it results in a relatively small engine because of the smooth and straight flow path.

The disadvantage of the method of increasing pressure and temperature lies in the fact that fuel consumption is reduced for full load, however, the gain for part load is only insignificant. In any case, this is a fundamental disadvantage of the constant-pressure gas turbine which also gave it the name "full load" machine. This expression is supposed to indicate that the engine shows its best characteristics at full load, especially the lowest fuel consumption. This characteristic points out the direction of the promising application of gas turbines, significant examples of which are propulsion of airplanes and ships for which output is high and constant for almost the whole operating time. In contrast, the load of rail cars and road vehicles varies considerably and often, and is generally low in relation to full load.

The conditions become more favorable when using a heat exchanger which requires a lower cycle pressure ratio. Utilization of the exhaust gas energy increases with increased heat exchanger efficiency. Two types of heat exchangers are available: the recuperator and the regenerator. In the first one, the cold air which is to be heated and the hot gas which is to furnish the heat are separated from each other by means of tubing, sheet-metal walls, etc. For the second, air and gas flow through a regenerator matrix which absorbs the heat of the hot gases and then transmits it to the cold air. For this purpose, the matrix is moved from the hot gas side to the cold air side, mostly by means of rotating a drum or disk which contains it ("rotary regenerator"). The advantages of the recuperator are the known relationships and designs available from other applications. However, the frequent temperature changes and the resulting heat expansions and stresses in a vehicular gas turbine have to be considered. The recuperator's main disadvantages are the large volume required for high efficiency and the requirements for stainless materials which jeopardize the advantages of low weight and space requirements for the whole engine. The recuperator is used extensively for intermediate- and high-power class applications. For stationary applications, there are further possibilities to improve the thermal efficiency: multi-stage compression with intermediate cooling, and multi-stage expansion with intermediate heating, as well as closed-cycle layouts.

Concerning weight and space requirements, the regenerator offers considerable advantages due to the application of narrow flow channels inside the regenerator matrix. The new problem concerning this type, however, is the sealing of the high-pressure portion toward the low-pressure one (atmosphere) which must be effected at the rotating drum or disk. The application of this type depends mainly on the resolution of this problem, the answer to which will considerably affect the specific characteristics of gas turbines, i.e., the lack of mechanically rubbing parts and the constancy of output and fuel consumption over long operating times, as well as the mechanical simplicity. Partial solutions were found after lengthy development and testing, however, the rotating heat exchanger can by no means be considered as an operational component.

Other Operating Processes

The concept of gas turbines is not limited to constant-pressure gas turbines as discussed above, but includes all units that contain any type of gas turbine engine. As is well known, the first commercial gas turbine engines, e.g., operated according to the constant-volume process with discontinuous combustion; this process entails considerable disadvantages and complications and, therefore, was unable to lead to any continued success.

For larger installations the possibility exists to gain advantages through the combination of steam and gas turbines; for intermediate and small engines the gas generator may be replaced by a piston engine which may also operate with the free-piston engine principle. This results in higher thermal efficiency; however, the advantages of a gas turbine constructed of only rotating parts are lost.

Summary and Conclusions

The above considerations show the development of the gas turbine for use other than aviation applications such as has emerged during the last several years after the initial turbulent planning; it also gives indications as to future development work.

While for some areas of application solid design principles and configurations have already crystallized, developments in other areas are still in progress. The latter is especially true of the very high and very low output classes.

The most favorable prospects for gas turbines are in the range from 1000 to 5000 hp. For this output class their specific characteristics for various applications as compared to Diesel engines and the steam turbines become especially favorable. For higher output the development time is inherently longer and exact comparisons with steam turbines of this capacity are not yet available. However, it is very likely that the conditions will be favorable for gas turbines as peak power producers. In the lower output classes the dominant position of the Otto and Diesel engines is not yet endangered by the gas turbine except in the cases where its high fuel consumption is acceptable in favor of its weight and space requirements and its other characteristics.

Besides, since all attempts to reduce the fuel consumption will result in increased weight and volume, abandonment of the mechanical simplicity and facility of inspection, as well as increased susceptibility and shorter overhaul intervals, it is in any case doubtful whether the gas turbine will be able to replace the piston engine in the low-power ranges. There are few characteristics of the gas turbine in this power class which cannot also be achieved by piston engines and auxiliaries; the gas turbine configuration leads to many problems in the manufacture, the materials, operation and installation, the solutions of which cannot yet be estimated. It will, therefore, be useful to follow the development of gas turbines in the lower range of output very carefully.

Chapter 7

Aircraft Gas Turbines, Revisited

Product Improvement Programs

My professional career of a quarter of a century in Europe continued on a hot July day in 1956 in the New World, full of hope and expectations for a rewarding future. In my first days at the Wright Aeronautical Division (WAD) of the Curtiss-Wright Corporation (CW), I experienced a few changes, to put it mildly. Lundquist had moved to the Corporate Office and Farrar suffered a fatal heart attack. So I was on my own, without my mentors. In the waiting period for my security clearance, I was as Staff Engineer of the Design Department put in charge of design for the Product Improvement Programs (P.I.P.) for the reciprocating and single-spool jet engines, as well as for the advancement of turbine cooling technology.

Impressive for me was the large number of engines in commercial and military service, their maturity, and reputation. The piston engines included the air-cooled single-row Cyclone engines with seven and nine cylinders, and the double-row with 18 cylinders. The power of the 1820 cu.in. Cyclone 9 had been raised from 575 hp in 1931 to 1525 hp in 1955. The 3350 cu.in. double-row had reached 2200 hp, then as TC18 Turbo Compound 3850 hp.[1] Over forty airlines flew Wright piston engines. One model was sublicensed for production. The Wright turbojets, J 65 Sapphire, an Americanized Armstrong-Siddeley engine, were powering fighters and fighter bombers of the U.S. and NATO countries. The U.S. Navy had chosen the supersonic F11F-1 "Tiger" for its Blue Angels demonstration team.

Compared to my work in Europe at Heinkel and Parsons, I found the annual P.I.P. budget enormous, both in manpower and money. I faced an entirely new situation, not in technological but in organizational terms. The lines of communication from the field, the commercial airlines and the military squadrons to the Design Department consisted almost exclusively of paper, reports, memos, requests, and so on. First I assumed that the sheer size of the organization necessitated this

---

[1] The highest power I had ever witnessed was a 14-cylinder double-row engine on a Bramo test stand producing 2000 hp.

cumbersome procedure. I reluctantly adjusted to it. Gradually, I discerned some flaws. Each department protected its turf. On the long route through the various channels, important details were being distorted or lost. I injected more realism into the system and brought the design engineers under my jurisdiction closer to the actual part, its operation and its mode of failure. Under this streamlined operation we acted as a team, with the result of achieving better product improvement in a shorter time.

Other items I also found new — and strange — included the fact that engines for the U.S. Air Force and Navy carried different model numbers. This I could understand due to their different accessories. But the engine components were occasionally also different which looked strange to me. The reason for this anomaly was simple: One service would accept a product improvement item, the other would not.

Inspecting an engine part I saw its inscribed part number but not WAD's trademark. In my opinion, the latter was necessary in order to trace bogus parts and their makers. My suggestion to add the trademark to each part as a legal identification was refused for two reasons: (1) it would cost money and (2) the origin of all parts can always be ascertained.[2]

One of my early P.I.P. assignments included the radial double-row engine Turbo Compound TC18. I am happy to state that the turbomachines themselves were operating perfectly in both commercial and military service. The turbine design, à la Moss, relied more on heat-resistant alloys than on internal cooling of the blades. A separate impeller supplied cooling air to the turbine disk through elongated holes in the blade roots to the other side of the disk and to a cooling shield with a duct to the exhaust gas stream. The outer blade portion was hollow but uncooled. The mixture of cooling air and exhaust gas burned when the engine operated with a rich fuel/air mixture at takeoff and climb. The resulting foot-long flames appeared frightening. In a pamphlet on the airplane and engine, passengers were calmed with the remark that these flames were normal and no cause for worry.

Other TC18 engine components caused some service problems. The two-planetary drive of the supercharger contained a small, highly stressed spur gear that spalled and cracked occasionally. A three-planetary drive as replacement carried a 38-pound weight penalty causing the airlines to reject it. This gear problem and that of the exhaust valve seat made me realize the difficulty of making an almost perfect component really perfect. Still, the TC18 Turbo Compound presented the *non plus ultra* of the high-power reciprocating aircraft engine. In 1957, this engine

---

[2] A few years later, bogus parts appeared in Wright engines. Then I was told that the ensuing millions of dollars litigation would have been much simpler if my suggestion had been put into effect.

permitted a regular round-the-world 28,435-mile six-stop route flown in 104 hours. The U.S. Navy achieved with surveillance aircraft on-the-wing operating times of 4800 hours.

My personal contributions to the P.I.P. program included: a cage for the J65 turbine roller bearing that withstood marginal lubrication; in the supercharger drive erosion elimination of a bronze bushing by Michelle-type introduction of the oil flow; and for direct-drive engines a hydraulic torque meter that provided accurate engine torque readings at all speeds and was insensitive to sludge in the oil. A patent was granted for this device.

Off-duty time was very pleasing. For winding down and sports, CW operated the Lake Rickabear Club, whose membership was confined to Managerial, Supervisory and Administrative employees, their immediate families, and guests. Its lake, sandy beach, forest and trails presented excellent facilities for swimming, boating, hiking, tennis and other games, ice speed and figure skating, tobagganing and cross-country skiing. And there was a cafeteria and picnic area. It was an ideal place for keeping fit and for relaxation which we, young and old, enjoyed in summer and winter. We appreciated that all, but food was free! Our move to the U.S. had financially suffered from the then-fixed exchange rate of 24 cents for a Deutschmark; it had forced us to a large first and a second mortgage on our home. Thus we could hardly afford an equivalent country club. Our do-it-yourself skills helped to stretch the budget, I as an engineer/handyman for modernization and repairs, and my wife as an amateur dress designer and seamstress for dresses for the girls and herself, as shown in the family photo (Figure 7.1). In short, we were an exception to the then prevailing affluent society!

Being a member of German professional societies, the Verein Deutscher Ingenieure (VDI, Society of German Engineers) and the Deutsche Gesellschaft für Luft- und Raumfahrt (DGLR, German Society for Aeronautics and Astronautics; successor to the Lilienthal Gesellschaft), I joined the SAE and was accepted as a full member in 1957.

The enthusiasm for my having joined Curtiss-Wright received a lift from several events. I was impressed by a lecture series presented by venerable scientists on acute subjects relevant to Curtiss-Wright's future. Two announcements indicated that this future looked promising: CW planned to build a new high-rise engineering building[3] and had reached agreements between Daimler-Benz, Studebaker-Packard and Curtiss-Wright providing for a fully integrated program of engineering, production, sales and service of automotive vehicles, to be extended to aircraft products.

---

[3] My joy in selecting a corner office on the sixth floor was subdued when an old-timer warned me, "I believe in this building project only when I see bulldozers preparing the ground for it."

139

Fig. 7.1. The Bentele Family in Ridgewood, NJ, 1956. Clockwise from top left: Max, Magda, Rose-Marie, Ursula and Brigitte.

At that time, the three companies together employed 107,000 people who generated total sales of $1,266,000,000. Curtiss-Wright was at the top with 32,000 employees and sales of $571,000,000.

Daimler-Benz (DB) and their Mercedes cars were not yet household words in the U.S. Just one example: At his Chicago hotel a DB executive asked the parking valet for his Mercedes 300 (then the largest model, with a 3-liter engine). Unable to recognize the make the valet wanted a description. He interrupted that with the words, "Now I know, the little one."

At a dinner we talked about automobiles, American and theirs, the best-engineered in the world but expensive. To our question of which car offers the greatest value, car for dollar, without hesitation all three answered in unison, "a Chevy!"

Another noteworthy anecdote: At a Stuttgart meeting with DB engineering executives, a CW V.P. and me in attendance, all discussion was conducted in English. Design engineers called in for sticky items, however, required translations which the DB execs provided. This back and forth talking went on for an appreciable length of time. Then I interrupted the translation of one of my points in Swabian dialect with, "Sorry, you must have misunderstood; that's not what I'd said!" With obvious consternation of all DB people I was asked, "You know German?" After my reply, "Of course. I was born on the Alb and had studied here on the T.H.," one could feel that they all were rerunning their mental tapes to find items they had discussed in Swabian which were not meant for us Americans to know. Caveat your own language and that of your opponent!

Market Life Extension

Dieselization

The status of the diesel program was easy to comprehend. Converting aero-engines to diesels presented some pitfalls that Curtiss-Wright considered severe enough to abandon plans for production. That decision was good luck in disguise. During the fuel crisis of the 1970s, the General Motors Corporation tried this dieselization approach with its Oldsmobile V8 automobile engine and marketed it as the only American mass-produced diesel engine for passenger cars. Many customer complaints, service problems and lawsuits finally forced GM to abandon this engine. The bright image of the diesel passenger car had been appreciably tarnished; as of this writing, the lawsuits are still going on.

Muffling

In the mid-1950s, the oncoming jetliners promised smoother flights and less cabin noise. In cooperation with the Douglas Aircraft Company (DAC), Curtiss-Wright initiated a program to investigate the effect of muffling the exhaust noise of the three TC18 turbines on the sound level inside a DC-7. Originally, the Research Division carried the program, code-named Project Smoothie. On account of my experience with Helmholtz resonator sound dampers, I was asked to participate and conduct the noise evaluation program on a Turbo Compound TC18 installed in the nose of a B-17 aircraft (see Figure 7.2(a) and (b)). DAC provided and monitored the microphone and recording equipment.

The measured effectiveness of the Curtiss-Wright Research Division mufflers markedly differed from the theoretical predictions, greatly disappointing its designers. My Helmholtz-type mufflers improved the attenuation characteristics considerably; unfortunately, further noise reduction in the desirable frequency range by fine-tuning of the resonators had to be abandoned for time and financial reasons. Though these preliminary mufflers did not reduce the overall noise level appreciably, they improved it to the more acceptable quality of a turboprop.

This program was my first excursion into Aeroacoustics; for reciprocating engines, it was obviously too late. It gives me great satisfaction to observe that Helmholtz resonators are receiving renewed attention for noise reduction in the aviation field.

Propellertrain

CW's aircraft engines also were considered for railroad use, a less obvious application. When I read in the *New York Times* an article in which Hurley proposed a Turbo Compound TC18/propeller system for pulling railroad cars, illustrated with an artist's conception, I held my breath. As an engineering student, I had been intrigued by the Schienen-Zeppelin (Zeppelin airship on rails). It ran as a shuttle between Berlin and Hamburg. Franz Kruckenberg had conceived and developed this type of railroad propulsion. An engine-driven propeller, neatly tucked into the upper corner of the last of streamlined railroad cars, pushed the train to high speeds. In 1931, the then-record of 230 km/h (143 mph) was achieved. This propulsion suffered from two major side effects, a technical and an environmental/societal one. The propeller had higher energy losses than mechanical, hydraulic or electric transmission means. This drawback did not compensate for its benefits, the low wear and tear of the wheel/track interface. The other effect was more subtle. The propeller acted as a vacuum cleaner and a powerful fan. Passengers waiting on the platform for a conventional train had to hold their garments when the speed shuttle passed through. Otherwise, ladies would be rather embarrassed by a scene which, two decades later, Marilyn Monroe made hilariously famous in the film "The Seven Year Itch" (1955).

For these reasons, the German railroad had to abandon the propeller in favor of driving through wheels. When I cautiously told Hurley this story, he ignored it

Fig. 7.2 (a), (b). TC18 exhaust muffling.

citing two factors, allegedly in his favor, i.e., more engine power and a reversible propeller for braking the train. He could not convince the American railroads, either. Apart from a granted patent, nothing became of this project.

High-speed trains still rely on driven wheels. In 1972, the Japanese bullet train Shinkunsen made history. In the 1980s, the French TGV (train à grand vitesse) established new speed records with 300 km/h (186 mph).

Parallel to Hurley's propeller-driven train, new concepts were proposed to eliminate the driven-wheel/rail system: air cushions and magnetic levitation (maglev). Curtiss-Wright investigated air-cushion vehicles, demonstrated one at the Rockefeller Center in New York City, and proposed an Air Car (see Figure 7.3).[4] Road holding, braking, among others, posed insurmountable difficulties and caused abandonment of the project.[5] Air-cushion vehicle technology achieved some success with military amphibious landing crafts and with passenger boats. Air-cushioning also was proposed for railroads, without success.

The maglev system, originally pursued in the U.S., is being developed in Japan and West Germany. Both countries operate experimental test tracks; a German Transrapid train achieved thereon a speed of 435 km/h (270 mph). To demonstrate the new technology, a link of the Düsseldorf and Cologne-Bonn airports is under construction. In competition, Japan also started a similar project. Other commercial applications appear to be imminent, also in the U.S. The competition between TGV and maglev is still very much alive: The French SNCF railroad exceeded the maglev record in May 1990 with a speed of 515 km/h (320 mph). I found it noteworthy to learn that the wheel/track system experiences chatter marks at high speeds which cause deterioration of the track. The Wankel rotary engine suffered initially from the same phenomenon. It was alleviated by a special material combination of the rotating apex seals and the stationary trochoid housing. It might be difficult to apply this solution to high-speed rail cars.

Though Curtiss-Wright's life-extension efforts lacked tangible results, they provided the engineering community with valuable lessons and thereby served a useful purpose for future ventures. We want to learn from history, don't we? I was glad to return to more promising work.

---

[4] I had the pleasure to drive this air-cushion vehicle; it was a unique experience. However, I was glad I was in a large empty parking lot; I would not have dared to drive it on a public street.

[5] It appeared useful for specific applications. As a result of the publicity of the NYC demonstration, a midwestern farmer had sent to CW a blank cashier's check for the purchase of such a vehicle. A river divided his huge farmland. With it he would be able to cross it — and save a lot of valuable time, he wrote.

CURTISS-WRIGHT 2-PASSENGER AIR-CAR

Fig. 7.3 (a). Curtiss-Wright air-car.

Here is the world's newest method of transportation...the Curtiss-Wright

AIR-CAR

Without the usual wheels, axles, transmission, clutches, etc., the Curtiss-Wright Air-Car travels smoothly over any unobstructed terrain, across water, mud or swamps on a cushion of low pressure, low velocity air. The controls are simple and the vehicle is inherently stable — anyone who can drive a car can operate an Air-Car.

The "Bee" is a compact, two passenger Air-Car which travels at speeds up to 56 mph with a 100 horsepower engine. This is the next model planned for production.

The Model 2500 is a 2-engine, 300 horsepower vehicle which comfortably carries four passengers.

Write for information to Dept. AC, Curtiss-Wright Corporation, 304 Valley Blvd., Wood-Ridge, N. J.

Fig. 7.3 (b). Curtiss-Wright air-car.

Turbine Cooling

Convection Cooling

At Heinkel-Hirth, I had successfully applied air-cooling to the turbines of turbochargers and jet engines. In 1945 I was interrogated by engineers of the Western Occupation Forces on this subject. Since then I had continued, on and off, to work on turbine cooling. One example is a patent for an air-cooled blade, applied for in 1952, assigned to Heinkel and granted in 1955.

With this background I was eager to obtain first-hand knowledge of the progress made with turbine-cooling technology in the U.S. Sorry to say, what I saw and read disappointed me; it indicated that progress in the last decade had been very slow, indeed.

The J65 engine, admittedly designed some time ago, incorporated uncooled solid turbine blades only. To improve the durability of the first-stage turbine stators, in addition to material improvement, a 150-hour test with hollow air-cooled blades had recently been completed with the statement, "Results encouraging, development to continue in 1957."

In June 1956 at a meeting in Cleveland, Ohio, the NACA Lewis Flight Propulsion Laboratory had suggested exercising caution with the application of air-cooling to gas turbines. It may, apart from the associated complications, increase specific fuel consumption of the engine.

An Armstrong-Siddeley forged rotor blade I was shown consisted of two halves brazed together, forming two cooling channels. A rudimentary analysis showed that cooling of the high-stressed portion of the blade and of the leading and trailing edges was insufficient. This cooled blade was, therefore, hardly superior to a solid one.

It appeared I was at the right place at the right time for advancing the state-of-the-art of turbine cooling.

The road to effectively cooled turbines included design, heat-transfer analysis and fabrication of forged and cast blades, design and manufacture of a cascade rig for their testing under simulated operating conditions, and finally engine testing of promising configurations.

At that time Curtiss-Wright operated rotary forging equipment that enabled the forming of tubes with variable wall thickness and with internal integral fins of different shapes. Feasibility studies and experiments with this method of producing hollow blades were abandoned, however, when superior cast blade materials became available.

146

Initially, cast blades had to be split and the two halves brazed together. This design provided great freedom for the inside configuration of the cooling passages. It allowed adjustment of the internal to the external heat flux at each blade section. The targets for the blade temperatures were well known and established as follows: chordwise a fairly uniform temperature distribution to avoid additional thermal stresses; lengthwise a temperature profile that provides sufficient safety margin for stress fracture. The amount of cooling air had to be consistent with the desirable engine performance and blade life.

To achieve these goals we employed high internal heat-transfer coefficients and large internal cooling surface areas. Specifically, our designs included: at mid-chord, straight unobstructed cooling channels; at the leading edge, additional fins and pins; and at the trailing edge, also pins. Exit of the cooling air was achieved through the trailing edge and/or on the pressure side of the airfoil upstream of the trailing edge.

Analytical and experimental results on the cascade rig with these blade designs were very satisfactory. Incorporating them in a turbofan engine would allow operating at an average turbine inlet temperature of 2000°F (1093°C) at both sea level and high altitude, and high Mach numbers. Since the designs had not been fully optimized, it was expected to extend the operating temperature to 2200°F (1204°C) and higher.

Further simplification was achieved with a cast one-piece blade. The internal configuration was formed by a machined core made from molybdenum that was leached out after casting.

With this air-cooling technology, we were able to replace the two turbine stages of the J65 engine with a single-stage design. It was, however, too late to phase it into the production of the engine.

The U.S. Navy considered our cooling technology the greatest advancement of the state-of-the-art. Regrettably, there was no active engine project at WAD in which to incorporate it.

Transpiration Cooling

Another Curtiss-Wright group concentrated its efforts exclusively on transpiration cooling of the turbine blades. I watched this competition but was confident that my cooling methods would prevail.

The transpiration-cooled blade consists of an airfoil-shaped strut covered with a thin sheet of porous material forming the external airfoil of the plate. The strut contains on its circumference a dozen narrow lengthwise channels; cooling air metered to these channels transpires through the porous material and mixes with the gas stream. The British are less euphemistic; they call this method sweat cooling.

147

The prime advantage of this cooling method lies in the effective cooling of the load-carrying strut. Strut and skin are protected from direct contact with the hot gas stream. However, the cooling air flowing through the porous sheet is injected at a right angle into the gas stream and, thus, detrimentally affects the aerodynamic performance of the blade. In contrast, internal convective cooling does not affect the external gas flow. On the contrary, judicious design of the cooling air exit can improve turbine performance.

Study Engines

Curtiss-Wright's success with the J65 jet engine, of which a very large number were produced and operated, lacked a follow-up. The next generation of high-performance jet engines required substantial increases of the cycle temperatures and pressure ratios. WAD had developed its own twin-spool machine for that purpose. Having no "need-to-know" yet, I was aware only of its model designation and its role in a dual-cycle propulsion system, a combination of a jet and a ramjet engine. The government contracts for full jet engine development to production were awarded to the competition: to GE for a single-spool configuration with a variable-stator compressor, and to Pratt & Whitney for a twin-spool machine. WAD's attempt to repeat the J65 success by Americanizing corresponding British Armstrong-Siddeley engines proved fruitless.

This drastic loss prompted layoffs and a reorganization at WAD. To explain the grave situation to the top echelons of the Division, our General Manager invited them for a pep talk to the auditorium, approximately 300 persons. He had hardly started to speak when a shop supervisor got up and asked, "Who are you?" This question illustrated WAD's type of management; it lacked the personal hands-on style.

I became Manager of Mechanical Components for advanced gas turbine engines, mainly compressors and cooled turbines. Efforts on cooled-turbine technology were expanded. Major work on compressors was dedicated to blade vibrations caused by forced excitation, rotating stall and aeroelastic instability, conventionally called self-excited flutter, or flutter for short.

I was familiar with forced excitations but had to brush up on the other phenomena.

Studies of aeroelastic problems date back to the late 1920s, mainly focusing on aircraft wings. Highly loaded compressors experienced blade vibrations and fatigue failures caused by rotating stall and self-excited flutter. The latter had spawned a number of theories and criteria worldwide. Quoted and applied were those of A.T.D.S. Carter, D.A. Kilpatrick, J.F. Shannon, among others. My study on the correlation of the design criteria, stability parameters and design constants for bending and torsional flutter with available test results revealed considerable controversies and the need for additional analytical and experimental investigations.

This challenging activity lost some of its luster when the USSR orbited its Sputnik in October 1957; the "Space Age" had started, and with it the emphasis on rocket rather than jet engine development. My own work was suddenly interrupted by the advent of a novel powerplant, a rotary engine.

Science to one is a goddess both heav'nly and high, - to another
Only an excellent cow, yielding the butter he wants.

Friedrich von Schiller (1759-1805)

Chapter 8

The Rotary Engine Era

The Wankel Engine

On a Friday afternoon in May 1958, Hurley called me to his office, handed me a crude plastic 5x6-1/2-inch model, and said, "That's supposed to be an engine. Find out over the weekend whether it is, its pros and cons. Don't talk to anyone about it. Report to me on Monday."

Full of anticipation, I showed the new and exciting contraption to my wife. She remarked, "Judging from its geometry, it looks elegant. If it works it could be great." Studying the "engine" in detail and step by step, I discovered that the arrangement represents an ingenious combination of the efficient four-stroke principle, the simple two-stroke valving mechanism, and a near-rotary motion. The engine offered low bulk and weight with additional advantages over reciprocating engines: more uniform torque, less vibration, and less sensitivity to the fuel octane number. Whether all of these advantages could be achieved in practice would depend on satisfactory solutions to its problem areas: gas and oil sealing, cooling, ignition and combustion. On the whole, I arrived at a positive conclusion.

On Monday I reported my findings in full detail to Hurley, beginning with "Yes, it's an engine" and concluding with "It's worthwhile to pursue it further." Only then did I learn that its inventor was Felix Wankel (1902-1988) and that NSU already experimented with it.[1]

My report was historic. From that seed stemmed the growth of the Wankel engine development in Germany and in the U.S., and Curtiss-Wright's envious position as exclusive licensor for North America and its participation in all fees and royalties from Wankel licensees wherever rotary engines are made and sold. The rest is history!

---

[1] NSU was a company producing bicycles, motorcycles, and small cars located at Neckarsulm, 50 miles north of Stuttgart. Its four-stroke-engined motorcycles were in the forefront of international racing.

151

The Wankel rotary engine development and its results are well-documented in a great number of papers presented at conferences of professional societies worldwide as well as in technical books. There is no need to repeat them here. The story behind this revolutionary engine received even more attention with articles in the trade and popular press and with books. Unfortunately, those publications were exclusively written by authors who had gained their information second-hand.

The allegedly definitive Wankel story was published in Germany in 1975 in a book entitled, <u>Der Wankel Motor: Protokoll einer Erfindung (The Wankel Engine: Protocol of an Invention)</u>. Its author had previously asked me some questions by mail which I had gladly answered. When I obtained a copy from a bookstore in Stuttgart, I almost fainted in disbelief reading this biased and distorted account of the actual developments. I wrote a book review for my own records (Appendix 8.1); it ends with the words, "Maybe I myself should write the definitive Wankel story one day. How about it? Let's hear from you." So I take this opportunity to describe some events as I experienced them.

Shortly after my report on the engine to Hurley and having received his clearance to give this information to his associates, I was asked to attend a meeting in a conference room. I waited there in anticipation. An elderly gentleman entered, said Hi, and told me, "We are going to fly to Germany to have a look at the engine." I politely interrupted with the question, "Do I know you?" After he recovered from this shock, he said, "In my many years at Curtiss-Wright, nobody ever dared to ask me such a question. You mean you really don't know who I am. My name is William L. Hanaway. I'm the corporate counsel reporting directly to President Hurley." I apologized for my ignorance. After this unconventional skirmish, we got along extremely well.

To prepare for my first encounter with the NSU management and engineers, I jotted on six sheets of paper my questions on the engine, its development status, problems, etc. On our visit to NSU, Hanaway was going to introduce me to the NSU president, Dr. Gerd Stieler von Heydekampf (1905-1983), who interrupted with, "Don't bother. We know Bentele very well. In the scooter years he was our fiercest competitor." The last word, "competitor," set the tone for our negotiations and cooperations.

I then met the engineers on the NSU/Wankel project, Dr. Walter Froede (1910-1984) and Ernst Hoeppner (1908-1966). They showed me all possible versions of rotary engines (see Figure 8.1). I was really surprised that the rotary engine I had deduced from the model was not the only one pursued by Wankel and NSU. There was a second one in which both the trochoid housing and the triangular rotor rotate. NSU had studied both types, the DKM (Drehkolbenmaschine, dual rotation machine) and the KKM (Kreiskolbenmaschine, stationary outer housing). The general public is familiar only with the KKM.

152

Wankel's original engine was the DKM-54, indicating the year of its invention. It is shown in Figure 8.2. The outer rotor, which contains the trochoid and the air intake passages in the side housings, rotates about its axis with a speed ratio of 3:2 relative to the inner rotor; that rotates about its own axis which is parallel with the distance of the eccentricity to that of the outer rotor. This arrangement offers only two advantages: (1) true rotary motion of both rotors and (2) unidirectional centrifugal forces on the gas seals of the inner rotor. The disadvantages, however, are substantial in the form of very complex systems for the gas flow, ignition, lubricating oil and cooling water. Still, it worked: I saw the DKM-54 running on the test stand!

The KKM, with the outer housing stationary, presents a much simpler engine, as shown in Figure 8.3. It depicts the first experimental engine designed and built at Curtiss-Wright. The only difficulty of this version is that the apex seals of the rotor are subject not only to outward centrifugal forces, as in the DKM, but on the trochoid's "ski jump" also to inward forces. Despite that hindrance, there was no doubt in my mind that we at Curtiss-Wright would consider only the KKM-type, and I made that position very clear.

Wankel's insistence and NSU's continual experimental work on the DKM indicated their preoccupation with high rotational speeds and the absence of a mechanical upper speed limit, which is a fairly common misconception of the characteristics of positive-displacement engines already mentioned. I also discerned that Wankel's ingenuity was apparently restricted to mechanical devices. His expertise in aerodynamics, thermodynamics and combustion was, to put it mildly, debatable. Examples such as the far-from-ideal combustion chamber shapes of his flat-disk valve and of his rotary engines substantiate that opinion.

Returning to WAD from NSU, I presented my findings and suggested taking an option for a license in order to secure time for further analytical, design and performance studies of the engine. NSU apparently misunderstood our careful approach and rejected our offer with the expression, "We don't take tips." Nevertheless, we continued investigating the engine's potential.

The results of our scrutiny of the patent situation in the U.S. led in July 1958 to another visit to NSU with Hanaway, *et. al.,* and me. Pointing at Hanaway, I told von Heydekampf, "If he had not made a gentleman's agreement, that we either acquire license rights from you, or we forget the engine, we would be able to develop and market the Wankel engine in the U.S. without you because you have no patent protection there." Von Heydekampf asked me, "You are not serious, are you?" I replied, "Yes, I am." Recovering from this shock, von Heydekampf immediately telephoned his patent attorney in West Berlin, told him his predicament and asked him to fly down post haste. To impress him with the seriousness of the situation, he, a stout 250-pound gentleman, was brought from the Stuttgart airport to NSU in

Fig. 8.1. Discussing the NSU/Wankel project were (from left) Dr. Walter Froede, myself, and Ernst Hoeppner.

a small car, an NSU Prinz. (It was a small car! Satirically it was said that one needed a shoehorn to get into it.) On his arrival, we conducted our heated, hours-long discussion in three languages, English, German, and Swabian dialect. Von Heydekampf was finally obliged to admit, "Bentele is right. We have no patent protection in the U.S."

In view of Curtiss-Wright's interest in the engine, we offered to assist in obtaining an adequate basic patent. Two French patents were in the way, one by Sensaud de Lavaud, which describes an engine similar to Wankel's DKM, and one by NSU itself.[2] We overcame both obstacles. Finally our combined diligent efforts and Curtiss-Wright's clout with the American Patent Office led to a basic U.S. patent of the Wankel engine.

---

[2] The French patent #853807, granted in 1940 to Dimitri Sensaud de Lavaud, describes rotary piston engines with parallel, fixed axes comprising two rotors that rotate with angular velocities inversely proportional to their number of lobes or teeth. One apparently advantageous form has convex inner walls of the outer rotor, which turns more slowly than the inner rotor. This may indicate that the other possibility — outer rotor faster than the inner rotor, i.e., Wankel's DKM geometry — was also known. NSU had carelessly allowed the issuance of its French patent prior to its application to the U.S. Patent Office.

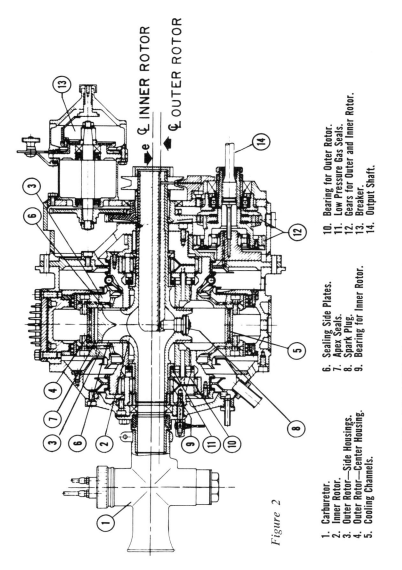

Figure 2

INNER ROTOR
OUTER ROTOR

1. Carburetor.
2. Inner Rotor.
3. Outer Rotor—Side Housings.
4. Outer Rotor—Center Housing.
5. Cooling Channels.

6. Sealing Side Plates.
7. Apex Seals.
8. Spark Plug.
9. Bearing for Inner Rotor.

10. Bearing for Outer Rotor.
11. Low Pressure Gas Seals.
12. Gears for Outer and Inner Rotor.
13. Breaker.
14. Output Shaft.

Fig. 8.2. NSU's first experimental DKM engine (1958).

By the middle of 1958, I had acquired a fairly good knowledge of the Wankel engine in comparison with other prime movers. I knew its pros and cons, so that when Hurley asked me whether I saw insurmountable problems, I said, "No, though the gas seals may take some time to develop." Hurley then asked me if I would be willing to head the R&D efforts for the rotary engine at Curtiss-Wright. I replied in the affirmative. A small office in an adjacent building, due to its color called the

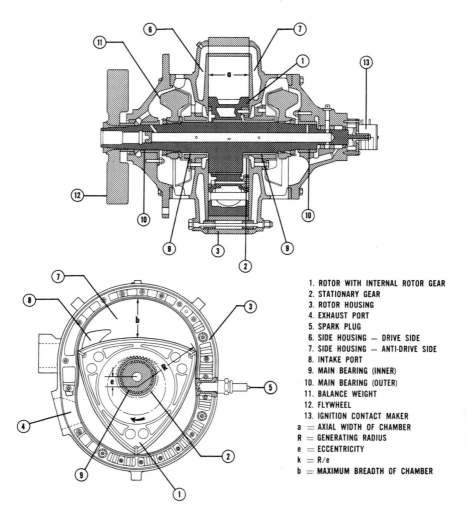

1. ROTOR WITH INTERNAL ROTOR GEAR
2. STATIONARY GEAR
3. ROTOR HOUSING
4. EXHAUST PORT
5. SPARK PLUG
6. SIDE HOUSING — DRIVE SIDE
7. SIDE HOUSING — ANTI-DRIVE SIDE
8. INTAKE PORT
9. MAIN BEARING (INNER)
10. MAIN BEARING (OUTER)
11. BALANCE WEIGHT
12. FLYWHEEL
13. IGNITION CONTACT MAKER
a = AXIAL WIDTH OF CHAMBER
R = GENERATING RADIUS
e = ECCENTRICITY
k = R/e
b = MAXIMUM BREADTH OF CHAMBER

Fig. 8.3. CW's first experimental engine (1958).

156

White House, was set up. I selected Charles (Charlie) Jones (1926-) and Ferdinand (Freddie) P. Sollinger (1896-1985), engineers I knew, for assisting me in the initial analysis, layout and design work.

We attacked with fundamental investigations the engine's problem areas, the structure and cooling of the housings and the rotor, the main bearings, and the oil and gas seals. The latter posed a dilemma. The NSU/Wankel gas sealing network was a Swiss-watch type, theoretically 100-percent effective but hardly practical and durable. Wankel being admired as the gas sealing expert, we were obliged to copy his design to the last detail for the time being.

Cooling of the trochoid housing presented the greatest challenge. In a reciprocating engine, cylinder head and piston dome are exposed to all phases of the working cycle. In contrast, the individual portions of the engine outer housing are exposed only to their respective phase. Temperature and flow velocity of the working medium, and thus the heat flux to the housing wall, vary along the trochoid circumference. Two extremes exist: In the combustion zone, the very high heat flux will induce nucleate boiling of the cooling water; in the intake area, the inflowing fuel/air mixture is, due to fuel evaporation, colder than the ambient atmosphere and much colder than the cooling water. The design task was therefore twofold: (1) to prevent overheating and resulting thermal cracking of the housing in the combustion zone and (2) to minimize thermal distortion of the housing and to assure effective gas seal operation.

We soon realized that NSU's circumferential cooling flow was incapable of fulfilling all requirements of high-performance engines. We selected axial flow, back and forth between the side housings, with the flow velocities in the various rotor housing sections adjusted to their respective heat input. In the combustion zone, the high-velocity cooling flow removes the steam bubbles generated by nucleate boiling, and thus permits extremely high heat fluxes to the housing without detrimental effects. This cooling scheme was also beneficial to the structural integrity of the rotor housing. Engine test runs confirmed its effectiveness in all respects. It was patented and became the standard for all high-performance Wankel engines.

NSU's experimental engines incorporated antifriction ball and roller bearings exclusively. According to our life prediction analyses, this bearing type was inadequate for our purpose. Consequently, we used sleeve bearings for the rotor and power shaft. (NSU also finally had to abandon antifriction bearings.)

In addition to these basic design investigations, we studied the porting and timing of the air and gas flow as well as geometrical effects such as the ratio of the trochoid generating radius to its eccentricity, the housing axial width relative to its axes, and other parameters.

After we felt comfortable with our preliminary selection of optimum engine geometries, I reported our progress to Hurley. He was pleased and ordered me to concentrate on the design of an experimental single-rotor engine delivering about 150 horsepower. Our performance calculations for this naturally aspirated engine led to a displacement of 60 cubic inches of *one* chamber; it was eight times larger than NSU's DKM and KKM engines. The specific power was 2.5 hp/cu.in.; in comparison, WAD's non-plus-ultra aero-engine, the Turbo Compound TC18, delivered at takeoff some 1.0 hp/cu.in.! We started the final design of America's first experimental Wankel engine at the end of September 1958; Figure 8.3 shows and explains its longitudinal and cross-sections.

The license agreement between Curtiss-Wright, Wankel and NSU was signed in October 1958. Curtiss-Wright obtained for a paltry down payment for the North American Continent the exclusive rights to all engine sizes and all applications.

Subsequently, my wife and I had the pleasure of receiving von Heydekampf in our home in Ridgewood. Reminiscing about the prior situation in Germany when he and I were competing in the two-wheeler market, he expressed satisfaction that we now cooperated on the same side. Turning to my wife, he said, "Believe me, the Wankel will bring revolution to the engine and automobile industries! All of us can look forward to a bright future!"

To get to that wonderland, my immediate, present time looked less pleasant. The research project I had first contemplated grew overnight into a major development program for the engine. Hurley envisaged three major projects for WAD. The first was to improve the production aircraft engines and to develop new jet engines. A second program was established for the above-described propeller train and a third for the Wankel engine, headed by me.

This reorganization put me in a very precarious position in two ways. I discerned a situation which in German is called *Tores-Schluss-Panik*. This phrase describes a panic due to concern that one is deprived of participating in a great historic event because the gate (*Tor*) is closed before one's eyes. In other words, expectations for quick results become extraordinarily great. Also, I was on the same or a higher organizational level as my former bosses. Although the pressure was tremendous, I was unable to refuse the assignment. It would have meant that I had misjudged the engine's potential and/or that I didn't trust my capabilities. To secure my continual services, "golden handcuffs" were laid on me: stock options and participation in Curtiss-Wright's incentive compensation program. (Each annual incentive compensation was spread over five years. If one decided to leave the company, one would lose the incentives awarded, but not yet paid out.) In short, I had no choice. I did my best under the circumstances. I also expanded my diary to include details of significant events.

I lost my effective "skunk work"; from one day to another, the engineers under my jurisdiction increased from four to four score. Apart from experimental engines, all imaginable engine applications were studied and pursued, on paper and often with installation mockups: boats, cars, buses, trucks and tractors, industrial drives and generator sets. A double-decker bus received priority; I liked the wooden mockup as a reminder of England, not as a viable engine application.

At that time, vertical takeoff and landing airplanes (VTOL) were under development and very much in the news. Canada had an airplane with tilt-wings with an engine and propeller at each end. Curtiss-Wright also had flown an airplane with two tilted propellers and had started a project for a larger airplane, the X-19, with four propellers. The engine weight is critical for all vertical takeoff airplanes. Therefore, Curtiss-Wright's VTOL project manager wheedled Hurley to consider also the Wankel engine for this application, an over-optimistic, extremely risky proposition.

We completed the test engine design by the end of December 1958. Having been frustrated with machining mistakes and useless parts in the past, I had a warning printed on the upper left corner of each drawing, "*If you don't know, ask!*"

Our experimental rotary engine ran for the first time on March 4, 1959, five months and ten days after the go-ahead for its design and procurement. This feat, which included the layout, design and detail shop drawings, patterns, castings, special machine tools, test stand preparation and instrumentation, etc., was accomplished because we enjoyed the best expeditor, Hurley himself.

Initially, the engine did not run properly, by any means, mainly as a result of the poor functioning of the gas seals. Replicas of Wankel's elaborate, sensitive sealing network proved unsuitable for this engine size. Against Wankel's and NSU's advice, we conceived, designed and built a much simpler sealing system. It led, after resolving some manufacturing difficulties, to very satisfactory results. Its basic design features have been adopted in all rotary engines since.

The road to effective and durable gas seals wasn't an easy one. I suggested seeking the help of companies who had relevant experience with piston engines. In October 1959 we were happy to receive representatives of the prestigious Perfect Circle Company to discuss our sealing problem, especially with regard to seal materials. Hurley, who had invited them, asked them to keep the engine secret, to which they agreed. After I started to sketch the engine and explain its workings, Arthur M. Brenneke (1910-1988), its Chief Engineer, interrupted me with, "That's the Wankel engine." He had seen its concept at the Volvo Company in Sweden at a time when NSU/Wankel were peddling the engine all over the world. And we all had believed that our project was not known outside our White House!

159

In view of this situation, Hurley felt it did not serve any purpose to keep our program secret. He arranged for a press conference to show the engine to the public, and particularly to impress Curtiss-Wright's shareholders. The local and national press reported Curtiss-Wright's new venture. After his presentation to the journalists and analysts, one of them took me aside and said, "Some people feel Hurley is a bum, but I think he is smart."

Exhilarated with his coup, Hurley promised me a champagne dinner if we would achieve another milestone. I told him that he had promised celebrations at other test results such as the first consistent engine run, achieving a brake mean effective pressure (BMEP) of 100 pounds per square inch (psi), a power output of 100 horsepower, etc. Consequently, I didn't believe that this would ever happen. He got my message, but didn't know whom to include and finally decided to invite each and every person connected with the Wankel engine project at Curtiss-Wright, a total of some 170 people. The champagne luncheon was held on the last day of 1959 at a local restaurant. Hurley stood at the door to the dining room, greeting everyone with a handshake and with thanks for the good efforts. To this day I regret not having had a camera to take a picture of this scene. Above his head was the restaurant's sign, "Watch Your Step."

NSU and Wankel were angered about Curtiss-Wright's premature announcement of the engine and wanted to reciprocate. They arranged for the Verein Deutscher Ingenieure (VDI, Society of German Engineers) to hold an engineering conference in the Deutsches Museum in Munich. It took place on January 19, 1960, with representatives from science, research and development, academia, government, institutes and industry in the audience, in all some 800 people, 26 from abroad. My absence was conspicuous. Hurley had stopped my trip to Munich because he was afraid I would meet too many friends who would pump me for details of our engine development program. It is noteworthy that the conference was opened with the remark that 63 years before, Rudolf Diesel (1858-1913) himself had unveiled his engine at the same place.

NSU issued for the occasion a colorful brochure on the NSU/Wankel engine with a significant title page. It depicts: 1813 a locomotive with a steam engine, 1867 the Lenoir engine, 1876 the Otto engine, 1897 the Diesel engine, and finally, 1959, the Wankel engine. Curtiss-Wright's work is mentioned in very complimentary terms, as an enthusiastic, highly experienced and extraordinarily powerful partner. It also said it was expected that Curtiss-Wright would start production of engines for certain applications in appropriate sizes in the very near future.

An experimental engine was demonstrated driving a water pump for fire-fighting; it failed to produce an enthusiastic response from the audience.

The German picture magazine, *Quick*, showered Felix Wankel with accolades: "The automobile, invented for the second time; the thirty years war of Felix Wankel, 1930 outsider, 1959 a victor; on the threshold to the assembly line."

NSU had provided me with tapes of the proceedings. It was fascinating to listen to the various engineers, scientists and officials. While emphasizing their contribution to the engine's success, almost all speakers joined the bandwagon. The industry itself exerted more caution; it took almost a year before Fichtel & Sachs signed the second license.[3] Ironically, it restricted the power range to 12 horsepower and excluded passenger car applications!

In view of the ensuing publications on the engine in engineering magazines, Hurley asked me to prepare a paper for the Society of Automotive Engineers. He would write the introduction and present the paper (and me) at a session of the Metropolitan Section. He did that on March 10, 1960. The paper was well received and subsequently included in the 1961 SAE Transactions under both Hurley's and my name.

Under the circumstances, we had made good progress. We had accumulated 2000 test hours in 15 months; one engine had delivered 124.5 horsepower. (With 160 horsepower the RC1-60 engine finally exceeded its design target of 150 hp; it became the workhorse for a wide variety of experiments.)

We were able to demonstrate an engine to the U.S. Army, had designed engines with more than one rotor, and actually built and tested an engine with four rotors. This RC4-60 engine ran in April 1960 and produced 425 horsepower. A marine version incorporating a gearbox is shown in Figure 8.4. It was planned to install two of these engines in a 48-foot 1000-horsepower boat. The engine's capabilities would be demonstrated on the Long Island Sound with Hurley at the helm passing all boats in sight at an unimaginably high speed. This historic event did not come to pass for various reasons. Hurley had tremendous success with Curtiss-Wright's mass-production of the "Anglo-American Jewel," the J65 jet engine, the Americanized derivative of the British Armstrong-Siddeley Sapphire. He was eager to add with the Wankel engine another triumph before his retirement (Tores-Schluss-Panik). However, Curtiss-Wright's fortunes had declined too far. At the end of May 1960, Hurley was forced to resign. Roland T. Berner (1909-1990), who had been a member of Curtiss-Wright's board of directors since 1949, was elected as his successor. Hurley was supposed to continue his relationship with Curtiss-Wright as a consultant. He told me, "I will be seeing even more of you now."

I harbored ambivalent feelings about Hurley. I admired his energy and drive, but he had no grasp of the R&D effort in time and money needed to bring the Wankel

---

[3] In 1958, CW had salvaged the Wankel engine with a small down payment for the first license.

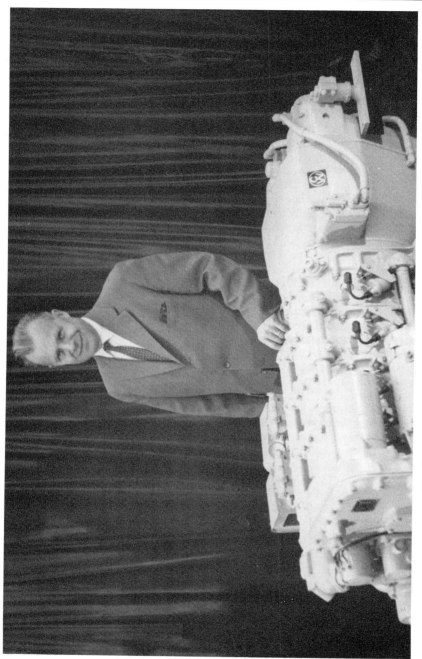

Fig. 8.4. Marine version of four-rotor RC4-60 engine (1960).

engine, a brand-new concept, to market. His vision had apparently been blurred by his successes in the Second World War and the Korean War. I also disliked that he accepted advice from his cronies of these times who promised unattainable production targets, rather than from R&D experts including mine, who carried the responsibility for the development of competitive powerplants.

After Berner had delivered to us his state-of-the-corporation message, one of my colleagues, an old-timer, said with a sigh of relief, "It feels like a breath of fresh air!" I kept my cool wondering what concrete plans for the rotary engine would emerge. The immediate events were hardly encouraging; they were more directed to the past than the future. Two words may describe them: witch hunt and retribution. After a few months in the breeze, my above-mentioned colleague left for another job.

CW's Corporate Office started to scrutinize the rotary engine test results as reported to the Board of Directors and published in the March SAE paper, assuming that they were subjected to *corriger la fortune*. Since that suspicion proved to be false, the SAE paper was to be ignored.

Hurley's services were frowned upon. He was unhappy and sued CW. For its defense, CW hired a prestigious New York law firm. One day I was ordered to cooperate with two of its representatives, open all files to them and answer all questions. One lawyer would lead the inquiry, the other would keep records. At the second meeting, I suggested to make a few observations, but only off the record. They agreed. I said, "Two things are on my mind. First, you are not only investigating Hurley but also me. Second, this suit will not go to court." They assured me I was mistaken with my first opinion and wished to know the basis for my second one. I explained it as follows: "I am only vaguely familiar with the laws applicable to stock corporations but I know this much. Hurley was president and chairman of the board, accountable to the board. The board determines corporate policy and as a watch dog has to monitor deviations, and stop them. In Hurley's case, the board did not diligently fulfill its obligation. The judge may ask the new management some questions in this respect. To eliminate this embarrassment, all parties will prefer an out-of-court settlement."

Checking more and more details of the official documents and my diary, the two lawyers were still unable to find a skeleton in Hurley's or my closet. After 17 sessions with them, they elaborately described my work for Hurley in a lengthy report for the CW management. Finally, the case was amicably settled. Hurley received somewhat less than he had sued for, CW paid the hired law firm from the difference, and I was exonerated and promoted to Chief Scientist.

In the course of the reorganization, the rotating combustion engine development was integrated in the WAD organization. My new assignment relieved me of all administrative duties for some 80 engineers such as suggestions for merit increases and promotions, job titles, vacation schedules, attendance of professional conferences, business trips, etc. I enjoyed this "loss of status" very much. As an example, the form "Personnel Transaction and Evaluation Notice" contained the item "LOYALTY (to company policies - acceptance of free enterprise system) Explain rating." In the past I often had problems arriving at a fair rating. The recent events would have made that even more difficult.

My new freedom provided me with many opportunities. I was keen to investigate the fundamental limitations of the engine design parameters, the size effects, and the means of advancing the engine, particularly with regard to burning heavy fuels, such as diesel oil.

There were no basic problems in scaling the engine; it behaved as a reciprocating engine. Experimental engines with 4.3 and with 1900 cu.in. displacement were built and tested, without any surprises. The small air-cooled engine was technologically a jewel, but costwise hardly competitive. An incident illustrates the last item. Our General Manager admired its rotor as a good piece of engineering. I agreed and said, "It poses only one problem. We have to reduce its manufacturing cost by 90%." (Fichtel & Sachs partly solved the same problem with its lawnmower engine; the bottom side of this vertical engine was without gas seals!)

The 1900 cu.in. engine contained gas seals which were biased for ease of manufacturing and as a result sealed poorly. Otherwise, the engine demonstrated the size, weight and vibration advantages as did the smaller sizes.

I directed my main activity to the capability of the engine to burn heavy fuels and still preserve its major advantages.

My writing and lecturing activities continued at the Rotary Engine session of the SAE International Congress in Detroit in January 1961. Though the largest room was made available, there was a SRO audience. Froede and I described in detail the development status of the Wankel rotary engine and took turns answering the host of questions.[4] It was a most satisfying event to me; it was my first formal, public speech in the U.S.! The proceedings including the discussions were subsequently recorded in the 1961 SAE Transactions.[5]

---

[4] Froede was slightly upset about my presentation. My slides were in color, his were black-and-white!

[5] A noteworthy item was the wear of the trochoid housing and the apex seals vis-a-vis that of the cylinder and piston rings. I am still proud of my answer: We agree that the sliding velocities of some rotating combustion engines are higher than conventional piston speeds, but it is well-known that the highest cylinder wear, and also probably ring wear, does not occur at the highest piston velocity but at the reversals of the pistons. In this respect, we have an advantage because "We do not stop." After many setbacks with NSU's early production engines, diligent development finally led to a combination of wear surfaces which proves my point.

Invitations for further presentations of CW's Wankel engine developments immediately followed. I talked at SAE Section Meetings in New York City, Milwaukee (home of Outboard Marine Corp., OMC), Twin City (John Deere territory), Baltimore, and at Ivy League and other colleges. I was particularly proud to give one of the Edwin G. Baetjer II Colloquia at Princeton University. In addition to SAE papers, I wrote articles for the *Motortechnische Zeitschrift* (MTZ) and *Diesel & Gas Turbine Progress*. These lectures and published papers served the purpose to familiarize the engineering and business communities with the engine, and to induce sublicenses.[6] There always were pertinent questions, but not once entirely negative comments. The concern of toxic exhaust emissions still was in the distant future. I gladly prepared technical brochures for WAD's P.R. department.

I also was able to accept an invitation by the Stevens Institute of Technology, Hoboken, New Jersey, to teach graduate courses on "Modern and Advanced Internal Combustion Engines." After 30 years, I was back at academia. It was more work than I anticipated but a valuable experience. I enjoyed my association with students and faculty members. Teaching, however, was not my prime profession.

The Society of the Sigma Xi elected me a member of the Stevens Chapter in 1964. The American Institute of Aeronautics and Astronautics admitted me to the grade of Associate Fellow in 1966, in recognition of my professional standing and outstanding contributions. I appreciated these honors immensely.

Multifuel Rotary Engines

All Wankel engine experiments were initially conducted with gasoline-type fuels. It was desirable to extend the range of fuels to heavier varieties, diesel oil and jet fuel.

The 1960 reorganization gave me time to study this multifuel aspect. I also knew that prestigious diesel companies in Germany and England were interested in the Wankel. I wanted to have a head start on them.

For my fundamental investigations of the combustion process in internal combustion engines, I used my experience with the Otto-Diesel engine and the available results of experimental stratified-charge and multifuel piston engines.

The classic Diesel compression-ignition principle requires a much higher compression ratio than that of an Otto engine, which poses no significant problems with reciprocating piston engines. The Wankel engine, however, loses its major advantages with raising its compression ratio. Size and weight for a given

---

[6] Regrettably, I learned about the signing of the OMC license by an article in the newspaper.

displacement increase, and specific power output decreases. In addition, mechanical, gas sealing and combustion difficulties occur, the latter as a result of the increased peripheral length of the combustion chamber shape and its gas seals. Difficulties for starting and for low speeds will also be encountered. These reasons prompted my full exploration of low compression ratios for all types of fuel. This approach offered the additional benefit of the available thousands of hours of operation experience with rotary engines of low compression ratio. It finally led me to a stratified-charge configuration with coordinated fuel injection and ignition. This combustion method had been used in the Hesselman engine of the late 1920s and in the Texaco Combustion Process (TCP), invented by Everett M. Barber (1909-1987) in the 1940s.

Those engines employ a circular air swirl in the cylinder into which fuel is injected near top dead center, immediately ignited with a spark plug, and then burned at the rate of its injection. The high-intensity air swirl, generated externally, is associated with aerodynamic losses that lower the volumetric efficiency and thus the maximum power output of the engine. In contrast, in the Wankel engine the motion of the rotor relative to the housing generates the desirable airflow pattern in the combustion chamber; additional induction losses are thereby avoided.

I was happy to have my internal report, entitled "Feasibility Studies and Concepts of Multifuel Rotating Combustion Engines," completed in July 1961. On my subsequent trip to Europe, on which my wife accompanied me, we enjoyed many gratifying meetings with old friends and with companies working on the Wankel engine.

One highlight was our most gracious reception by Felix Wankel and his wife. One incident in their summer house on the Lake of Constance illustrated his excitement and state of mind. His representatives had returned from Japan with two license agreements, one (the third one) with Toyo Kogyo Co. (now the Mazda Motor Corp.) and one (the fourth) with the Yanmar Diesel Co. Ltd. He showed us the beautiful Japanese dolls he had received as goodwill presents. Then he said, "Almost every Japanese farmhouse is adorned with a picture of the Emperor and one of Rudolf Diesel, whose engine makes work easier for the farmer. Isn't it a grateful gesture to Diesel." The way he admired the dolls and he told the story gave to my wife and me the impression that he had the hope and wish that, in Japan, his picture would soon be added or replace that of Diesel. The enjoyable encounter was concluded with picture-taking (Figure 8.5).

166

Fig. 8.5. At Wankel's summer house. (From left) Myself, Mrs. Wankel, Magda, and Felix Wankel (1961).

Wankel also accompanied us to his new institute in its construction phase. It was in due course to be opened with a ceremony to which I would be invited, he said. I thanked him for his generosity remarking I was looking forward to this festive occasion.[7]

During our further stay in Europe, I found all engine companies pursuing the classic Diesel approach with a high compression ratio. On my question whether that will be successful, one reply was, "Yes, if one does it right." This statement indicated to me they were climbing the wrong tree. I was relieved to know that I was first with my configuration for a multifuel rotary engine, which became known as DISC (direct injection stratified charge). However, at a visit with Ricardo Consulting Engineers in England, I was cautioned that burning of heavy fuels definitely requires the high compression ratio of a Diesel. To my reply that Texaco engines operate with a relatively low compression ratio, I was told that their running hours were not long enough. In time they would incur coking, ring sticking, and other problems.

On our return, my family and I completed the final stages of the naturalization process at the end of 1961: An interview with a judge, the oath of allegiance at a solemn ceremony, and the receipt of the certificates of our citizenship of the United States of America.

Curtiss-Wright's experimental DISC engines, designed to my proposals, proved the viability of my concept. Various SAE and other papers describe these experiments in detail. In 1966, I was granted a basic patent on this engine, assigned to Curtiss-Wright (U.S. #3,246,636). Its preamble delineates the fundamental advantages that this type of Wankel engine offers for the burning of a wide range of fuel with this combustion process.

The rotary engine development was thus progressing in the proper direction and I was able to again return to the gas turbine field.

---

[7] This invitation never came. Furthermore, on a trip to Germany in 1973, I informed his office in advance of my wish to meet him. After many delays and excuses I finally received the message "Herr Dr. Wankel is unable to see you." I had become *persona non grata*, for reasons unknown to me.

Appendix 8.1

Book Review

The present lull in the Wankel engine hardware development is being utilized by engineer/journalists to publish THE STORY of this revolutionary engine. From Germany comes a new addition to this series, a lushly illustrated 222-page 9 by 11 inches coffee-table volume for 48 Deutschmarks. It is

Protokoll einer Erfindung: Der Wankelmotor
(Protocol of an Invention: The Wankel Engine)

According to its dust cover, "the author's prolonged contacts with the participants and developing companies, especially with Felix Wankel himself, gave him the possibilities to delineate the authentic history of an unusual invention." Too bad that this opportunity was missed on many occasions. Also, the author limited his personal contacts to the Wankel people within driving distance from his home near Stuttgart. The book should therefore by subtitled: "The authentic Wankel story — from the German point of view." Being familiar with some of the events and finding their illumination wanting and distorted (to put it mildly), I would even go a step further and suggest as the book's subtitle:

The glorious German Wankel team — and the ugly American(s).

Maybe I myself should write the definitive Wankel story one day. How about it? Let's hear from you.

Fairfield, October 18, 1975   M.B.

Chapter 9

Lightweight Aircraft Gas Turbine Engines

Lift/Cruise Engines

After the Sputnik shock of October 1957, the American space exploration efforts enjoyed priority. WAD conducted jet engine development with emphasis on lightweight, lift-cruise jet engines and turbofans. The analytical and design investigations finally led to a lift or cruise turbojet designated TJ60. By that time, jet and fan engines had regained government interest, and WAD was in the process of submitting a proposal to the U.S. Air Force for a joint development of this engine. I was asked to study the engine layout and comment on it.

As a lift engine, a target of a high thrust-weight ratio of 20:1 had been established; this number was 2.8:1 for the J65 engine! To achieve this high ratio, the compressor blading was designed with short chords and very high aspect ratios, i.e., the ratio of the length of the blade to its chord at midspan. Whereas the aspect ratio was conventionally on the order of 2.5 and 3.5, the aspect ratio of the front stages exceeded this number with values of 8 and more.

In addition to the centrifugal and gas bending loads, compressor blades are subject to vibratory stresses resulting from forced excitations and flutter. For their aeroelastic stability, the high aspect ratio, highly flexible blades require mechanical damping. Due to the small blade thickness, damping can be applied only to the root or to the tip of the blade. To prevent flutter, the larger fan blades are often equipped with integral snubbers. Their detrimental effect on aerodynamic performance is small and acceptable on fans, but not on compressors.

WAD used new design criteria expressed in stability diagrams whose accuracy had allegedly been experimentally confirmed; these were available for designing the proper dampers, considering both bending and torsional flutter. A variety of friction dampers for the blade tip and blade root had been analyzed and designed.

I expressed my concern and reservations on the engine layout, particularly on the high-aspect ratio compressor for two reasons, technological and organizational: (1) I intuitively believed such highly flexible blades would be unable to cope with all possible vibratory excitations under all operating flight conditions; friction damping at the upper and/or lower end of the blade may well be insufficient to reduce vibratory stresses to acceptable levels. The available experimental data could not convince me that blade stability could be permanently achieved. (2) To the best of my knowledge, all other investigators and companies apparently shied away from high aspect ratio blades. I disliked the single-track, high-risk mentality.

After the U.S. Air Force had accepted WAD's proposal and a competition for the engine for an American supersonic transport aircraft (SST) was announced, WAD formed a Task Force for these two projects. I became a member with responsibility for the mechanical design of the engines. This assignment was a welcome privilege. As head of a highly qualified, dedicated team comprising design, stress, vibration and heat-transfer engineers, I was eager to assist in achieving Berner's aim, namely to become engine company #1 again. Still, I felt uncomfortable in view of the skimpy available experimental database for the compressor and the transpiration-cooled turbine blades as well as the missing in-house test facilities.

Supersonic Transport Engine

WAD's most ambitious project ever was its supersonic transport engine for the American SST. Designated TJ70, it was based on the previously designed TJ60 engine, whose components were being readied for experimental evaluation. Though the air massflow of the TJ70 SST engine was five times larger than that of the TJ60, the associated scaling effects were considered to be in favor of the larger engine.

The TJ70 engine consisted of a seven-stage compressor with high aspect ratio blading of the front stages, an annular combustor with vaporizer tubes, a one-stage turbine with transpiration-cooled stator and rotor blades, a small afterburner, and a convergent-divergent exhaust nozzle. The structure of the airplane benefited from the engine's low weight; this positive feature was much appreciated by the airplane designers.

The Federal Aviation Administration (FAA), the airplane companies and the airlines expressed concern about the radical engine design and its lack of experimental substantiation. Specifically, the thin compressor blades, the effectiveness of their friction dampers, and their long-term durability were considered question marks. Further controversy related to the practicality of transpiration-cooling and to the safety of the two-bearing engine rotor in case of turbine blade failures. We were able to substantiate the two-bearing feature with our analysis and by citing successful flight operation of two-bearing engines of the past. Answers to the other items appeared less convincing. Confidence in paper or in running hardware became the crucial factor. WAD's lack of its own test facilities for compressors and other components, as well as insufficient test time with similar smaller engines, exacerbated the practical experience situation. In addition to our design efforts, Curtiss-Wright conducted a P.R. campaign to foster its SST competition. A small desk model of the TJ70 (Figure 9.1) illustrated its major feature, the short compressor.

The FAA arranged a conference held in January 1964 in a large auditorium in Washington, D.C. The three competing companies, GE, Pratt & Whitney, and Curtiss-Wright, made separate presentations of their engines to selected government agencies, the competing aircraft companies (Boeing, Lockheed and General Dynamics), and the major airlines. In the question-and-answer period after Curtiss-Wright's formal presentation, most questions that we had heard before came up again. I did the best I could to explain to the distinguished audience the compressor blading with its friction dampers, their effectiveness, and their long life.

On a mission in May of that year to Germany to "sell" Curtiss-Wright's lightweight engine technology to the German Ministry of Defense, I heard similar questions to those at the FAA conference. Again I did my best, which was apparently hardly good enough. This fact was illustrated on my visit to my parents-in-law in Heilbronn; they handed me a local newspaper with the headline, "Curtiss-Wright out of the SST competition."

*Fig. 9.1. Desk model of the Curtiss-Wright TJ70
supersonic transport engine (1963).*

173

GE won the SST engine development contract, mainly as a result of its vast experience with its J79 engine in supersonic fighters and bombers, and its J93 in the B-70 bomber. Four of its GE-4 SST engines were built and run successfully. GE's Vice President for Aircraft Gas Turbines, Gerhard Neumann (1917-) "Herman the German," was acclaimed and congratulated for this achievement. The U.S. Congress, however, finally killed the American SST Project in 1971.

The controversies in 1976 over the American landing rights of the British/French SST Concorde induced me to present an installment on the subject of supersonic and hypersonic flight to the Transportation History Foundation of the University of Wyoming. The list of the 18 folders and the flight milestones are included in Appendices 9.1 and 9.2.

Demonstration Engines

Concurrently with the SST competition, component and rig engine tests with advanced engine hardware achieved some excellent individual results.

The TJ60 rig compressor incorporated a rotor with high-aspect ratio blades and tip dampers for the first three stages; standard cantilever blades for stages 4 through 6; and stage 7 with a flexible shank. The stators for the first three stages were variable, and for the others adjustable.

WAD's newly developed design procedure was verified by the performance of a single-stage compressor. With a pressure ratio of 1.2, adiabatic efficiencies over 90%, with 1.5 over 85% were achieved. The effectiveness of tip dampers was successfully demonstrated with a two-stage turbofan. Tip dampers also were tested in vibration rigs. "No damper" blades failed after 15,000 cycles; "damper" blades achieved 10,000,000 cycles without cracks or failures.

The combustion system was tested in a combustor sector rig with operating temperatures up to 2500°F (1371°C); good temperature distribution factors and temperature exit profiles were obtained.

Transpiration-cooling was investigated with a variable-area turbine (à la Heinkel's He S 30). The nozzle blades are trunnion-mounted to allow approximately 20% area variation. A unison ring to which each blade is connected by a self-aligning crank is operated by two bell cranks positioned by an external actuator. Compressor air cools the nozzle blades and rotor blades and also the variable nozzle mechanism and support. This arrangement was tested in a sector rig and also as a full turbine in a J65 rig engine. The rig and rig engine were operated with gas temperatures up to 2500°F. Total engine test time of the rig engine was approximately 70 hours. I questioned the turbine efficiency, though.

174

These test results were interpreted to mean that a thrust weight ratio of 16:1 was in the cards.

The TJ60 rig engine was, however, plagued with compressor blade failures and other malfunctions. The predicted performance could not be demonstrated before the program was cancelled in the Spring of 1966.

The loss of government support for WAD's lightweight gas turbine engines gave me an ambivalent feeling. My dire concern and predictions about the flexible compressor blades and the transpiration-cooling became realities, which boosted my professional pride. However, they also boded ill for my future at WAD. This possible misfortune became clearer when I learned that GE had increased the chord of their compressor blades in the Summer of 1964, and that in January 1965 the U.S. Air Force already had established upper limits for the aspect ratio of blades in relation to their pressure ratio. Both facts had apparently been known to WAD program managers, but had been ignored. In the 1980s, the trend to low-aspect-ratio, cleaver-type blades continued.

WAD's type of transpiration-cooling had been found to cause appreciable downstream wakes resulting in an overall loss of turbine efficiency compared to convectively and partially film-cooled blades. Improved transpiration-cooling with air and steam now is being investigated and employed in industrial turbines.

Exotica

Liquid Metal Regenerator

WAD's preoccupation with uniqueness such as compressor blades with friction dampers and transpiration-cooled turbine blades was also extended to other gas turbine components and propulsion systems.

Heat exchange from the exhaust gas to the compressed air had been experimentally utilized for turboprop engines, with some increase in fuel economy. It is penalized with complexity of the engine air and gas flow systems and their associated aerodynamic losses. A liquid metal regenerator (LMR) can ease these conditions; its development was pursued within Curtiss-Wright.

An LMR consists of two sections: a "hot" section in the exhaust gas stream and a "cold" section located after the compressor. Either section consists of a set of finned tubes that contain the liquid metal, usually a mixture of sodium and potassium, called NaK. A pump circulates the liquid metal between the sections.

Those heat exchangers offer great improvements in the thermodynamic cycle, but these advantages are offset by operational and safety problems. NaK, solid at

175

room temperature, becomes liquid at temperatures below the boiling point of water; it also is heavily reactive with air and water. Any NaK leakage from the closed LM system can cause fires and lead to catastrophes.

In a competition for a turboprop engine incorporating heat exchange to increase the range of military transport aircraft, WAD's proposal was based on an LMR. I was in charge of the mechanical design section of the proposal. Considering any regenerated turboprop as a dead-end street, I was actually delighted when the G.M. Allison Division obtained the development contract with a conventional air-to-gas recuperator in the Summer of 1963.

In mid-1965, the U.S. Army Tank Automotive Command (TACOM) tried to advance the vehicular gas turbine technology by issuing an RFQ for a 1500-horsepower tank gas turbine with stringent requirements for the specific fuel consumption at three discrete power settings. All bidders — WAD included — met these figures in their respective proposals, otherwise their chance for a contract might have been minuscule. To fulfill the stipulated fuel consumptions, two options were available: either highly efficient turbomachinery and a heat exchange with a conventional recuperator, or a highly effective heat exchanger (like an LMR) and simple turbomachinery.[1] A radial-flow compressor, for example, also would be less sensitive to dust in the inlet air.

Single-Rotor Lift Fan Engine

The quest for vertical takeoff and landing aircraft (VTOL) had spawned a variety of schemes to provide lift for takeoff and landing. One concept envisaged in each of the two aircraft wings a large single-stage fan. A gas generator installed in the fuselage supplied gas to turbine blades attached to the fan's outer rim, and drove the fan rotor. GE operated such a configuration in a Ryan experimental VTOL.

WAD went a step further and eliminated the separate gas generator. The resulting single-rotor fan engine consisted of a large-diameter fan whose perimeter carried a combined gas generator/turbine. It was similar to the Lorenzen Exhaust Turbocharger of the late 1920s, with a combustor added. The self-contained engine looked very attractive, indeed. However, it posed the same severe air and gas seal problems which, among other drawbacks, had led to the demise of the Lorenzen turbocharger. In addition, the aerothermodynamic, design, reliability and safety

---

[1] Since WAD lacked experience with any other but a liquid metal regenerator, a tank engine was proposed which incorporated the latter. AVCO Lycoming had adopted a recuperator and obtained the development contract. At a debriefing session, the Army abstained from citing the fire hazard that an LMR would pose for the engine compartment. I believe, however, that it added to the items held against WAD. To the best of my knowledge, application of any type of heat exchanger to an aircraft engine has yet to make it to production.

problems were a formidable challenge to the design engineer. In my opinion, the difficulties were insurmountable in many if not all respects. Some high officials placed great hopes in this unique engine for VTOL aircraft.

On the day the Allison regenerative turboprop engine development was scuttled, I expressed to the WAD president my pride in having already recognized this fate at the time we had lost that competition to Allison. He replied, "That's easy for you to claim now." I countered, "I make now another prediction. Please write it in your notebook. WAD will never, and please underline never, sell its single-rotor fan engine." Both the regenerative turboprop and the single-rotor fan went quietly into oblivion, q.e.d.

Toroidal Drive

WAD also pursued other exotica. High hopes were placed on a toroidal traction drive for automotive and industrial applications. An outside engineer claimed to be able to eliminate the problems which had led to failures of previous designs, e.g., the Hayes transmission installed in cars by the Austin Co. in Great Britain, among others. Only peripherally involved in the project, I drew attention to the limited success of the Beier transmission in industrial applications and its inability to penetrate the automotive market. My warning was ignored. After some demonstrations in test cars and one industrial application the project had to be abandoned.

Miscellaneous

WAD pursued a variety of design studies and experiments for aviation and aerospace applications.

A multiplicity of VTOL nozzle concepts finally led to a simple thrust vectoring nozzle. A demonstration was conducted on a full-scale model.

A "diluent stator" was conceived for high-temperature gas turbines. It combined functions of the compressor air in the design of the first turbine nozzle: dilution of the combustion gases to the desired average turbine inlet temperature and its radial profile, as well as the cooling of the stator walls and structure; it also reduced the axial length of the combustion system. I am unaware whether it was ever installed in an engine.

A "cap pistol," to be used by astronauts for maneuvering during space walks, was designed and tested. Though this scheme offered many advantages, NASA questioned whether it was capable of providing the absolute reliability necessary for this purpose.

Subsequent duties also led me to the assessment of other corporate projects, mainly related to VTOL aircraft. Curtiss-Wright had pioneered the use of composites for hollow propeller blades. It offered weight savings and also aerodynamic shapes that were more effective for vertical lift. The blade skin consisted of fiberglass, the filler of foam. Such propeller blades were installed in the Canadair CL84 tilt-wing aircraft with two four-bladed propellers, as well as in the Curtiss-Wright X-2 with two and the X-19 with four tilt-propellers.

To eliminate or minimize plate vibrations of the skin, the blade pressure and suction sides were connected with numerous strings. This method appeared to me rather crude, but it served its purpose, at least for the short durations of experimental flights.

In the Summer of 1965, the X-19 lost a propeller during transition from hover to forward flight and crashed. The cause was traced to a fatigue failure of the gearbox housing, not to the composite propeller. Unfortunately, this accident led to the stoppage of the project.

In the mid-1980s, the X-19 aircraft type was revived with the Vertol/Bell X-22 Osprey. It experienced difficulties in obtaining funding for military and commercial applications.

Rolls-Royce introduced composite fan blades in its high-bypass turbofan engine RB 211 for the Lockheed L-1011 Tristar jetliner. This application was apparently somewhat premature. The composite material delaminated, forced an extensive and expensive redesign of the engine, and finally Rolls-Royce into bankruptcy in 1971. This calamity placed the U.S. in a dilemma. On the one hand, a formidable jet engine competitor was down; on the other hand, Lockheed's airplane was going with it. An AVCO division supplied wings for that plane; together with other interested parties it lobbied for a government loan guarantee to Lockheed. Congressional approval then rescued the Rolls-Royce aircraft engine division enabling it to redesign the engine and deliver it to Lockheed.

I imagine that Curtiss-Wright felt fortunate in having discontinued its composite propeller blade development after the X-19 crash. It might have caused similar difficulties and humiliations which Rolls-Royce had to experience.

An unexpected visit brought new perspectives to my activities at WAD. Siegfried Günter, my friend from Heinkel, stopped by when I was working on an engine proposal for the U.S. Air Force. He described the procedure used in the USSR to select an engine for development and production. The Soviet Ministry of Defense conducts a competition in which three factories are asked for a separate proposal. After studying and evaluating these, a conference would be held to which the participants would be chosen by the Ministry, not by the management of the factories involved. First, each factory team would present its engine in detail. Then the representatives of each factory would comment on the engine design of the two others, pointing out its strengths and weaknesses. Finally, the Ministry would combine the best features of the individual engine designs into the engine to be developed and produced. This "optimization" made sense to me.

Some twenty years later, the U.S. and the Western nations started to adopt a similar method. Development and production teams, combining resources of national and foreign companies, are in vogue now. These liaisons and cooperations radically change the aircraft industries worldwide in an ongoing process.

Transition

In 1966, Curtiss-Wright continued and intensified its restructuring. It favored a role as major supplier of components to the aerospace and other industries in lieu of its own gas turbine engines and power systems. Its letterhead omitted the once proud phrase "The World's Finest Aircraft Engines."

One bright spot was the Second Annual Rotating Combustion Conference, held at WAD in March of that year. It was attended by representatives from ten licensee companies: two from the U.S., six from Europe and two from Japan. (See Figure 9.2; WAD, the host, is flanked by the Japanese on the left, and the others on the right-hand side of the table.) I had managed to also invite suppliers of rotary engine components, Mahle (housings and their surface treatment) and Goetze A.G. (sealing elements).

The conference agenda and proceedings clearly established Curtiss-Wright's progress and success with the basic engine as well as its multifuel version. I was happy to be recognized for my fundamental engine development work. Most licensees had adopted our layouts and patents for gas sealing, rotor and housing cooling, as well as surface treatment of the trochoid.

In my *ad hoc* after-dinner speech at the conference closing, I recalled our early rotary engine days, triumphs and frustrations, and our final development successes. Though on occasion I had wondered about my 1958 assessment of the engine, this conference confirmed that it was the right one.

My ideas on the course of a novel engine from its invention through development to marketing and selling are presented herewith.

Fig. 9.2. The Second Annual Rotating Combustion Conference held at WAD in March 1966.

The introduction of a novel engine concept to the professional community and to the general public usually occurs in two phases: (1) a demonstration that the engine principle is sound and (2) a first extremely attractive application that will assure success in the marketplace.

The first phase requires an enthusiastic sponsor. Nikolaus August Otto had found it in the Deutz Company, Rudolf Diesel in Heinrich von Buz of the MAN Company, Hans von Ohain in Ernst Heinkel, and Felix Wankel first in G. Stieler von Heydekampf of NSU and then, after my positive assessment, in Roy T. Hurley of Curtiss-Wright. Frank Whittle lacked that sponsor; it cost him the prestige of being first with a jet-propelled aircraft flight.

The second phase posed no problem for the Otto, Diesel and jet engine; they all were uniquely capable of achievements unattainable before. That was not the case for the Wankel engine. Its distinct but not overwhelming advantages had to be clearly established in the first application.

The demonstration at the January 1960 Munich Symposium with a Wankel engine driving a fire pump was received with little enthusiasm. The same applies to Curtiss-Wright's first disclosure at the November 1959 press conference.

The first application of the new, revolutionary Wankel engine had to overcome this lethargy and present it as vastly superior to existing powerplants. In my humble opinion, that target was badly missed.

On one of my trips to NSU, an experimental single-rotor KKM was installed in a Prinz car chassis. My NSU counterparts invited me to drive the car on their test track. After a few runs I almost repeated to them my "it moved" statement made to Beier on the Volkswagen with his CVT. I was expected to praise the vibration-free engine operation. I simply replied, "First, if you are used to an American V8-powered sedan, you are not bothered and don't expect vibration problems. And, second, a single-rotor engine is in its instantaneous torque characteristic better than that of a single-cylinder four-stroke, but not much better than that of a two-stroke. The translational vibrations are eliminated but not the torsional ones." Unfortunately, my message did not sink in.

The first application of real importance was in 1964 with the NSU Spider, a two-seat sports car driven by a single-rotor KKM engine. The engine was installed in the rear, the radiator in the front. This installation failed to show the major advantages of the Wankel engine. When I asked von Heydekampf, "Why not a twin-rotor?", he simply answered, "A flywheel is cheaper than a rotor." My assessment of this first application proved finally right. The car did very well at some racing events but the buying public showed scant interest; only a few thousand of these cars were produced.

Hurley's plan of two four-rotor engines propelling a cruiser with raceboat speeds on the Long Island Sound would have been more impressive. It never materialized. None of Curtiss-Wright's numerous demonstrations in a car, truck, small boat, generator set, airplane and helicopter led to production of the engine, either.

CW's top management relished its leading position in rotary engine technology, but failed to exploit it, to concentrate on a specific application and to prepare for production of an engine model for it. This wait-and-see attitude brought sublicenses but did not create work for Curtiss-Wright's factories and employees.

The prospects for my usefulness to Curtiss-Wright had thus greatly diminished. These circumstances led to my amicable separation in May 1967. On several occasions, AVCO Lycoming had indicated to me that I should knock on its door if I was ever considering a change of employment. After a survey of the relevant American industry, I finally decided to accept the AVCO Lycoming offer of Consulting Engineer to the Vice President of Engineering. My first assignment consisted of the study and appraisal of another type of a rotary engine.

Appendix 9.1

MB Collection in Transportation History Foundation of the University of Wyoming

Folders:

 1.0 Supersonic and Hypersonic Flight

 1.1 Research and Flight Milestones

 1.2 U.S. Military

 1.3 Commercial (SST)

 1.4 Environment, Exhaust Emissions

 1.5 Sonic Boom, Noise

 2.0 Supersonic Military Aircraft

 2.1 American

 2.2 European

 3.0 American Supersonic Transport (SST)

 3.1 Development Program

 3.2 Boeing

 3.3 Lockheed

 4.0 French/British Supersonic Transport

 4.1 Concorde Development and Sales Program

 4.2 Concorde - Technical

 5.0 Russian SST Program

 6.0 Jet Engines for Supersonic Aircraft

 6.1 American Engine Developments

 6.2 General Electric (GE)

6.3   Pratt & Whitney Aircraft (P&W)

6.4   Curtiss-Wright (CW)

6.5   Bristol-Siddeley/Rolls-Royce/SNECMA

Books:

G. Geoffrey Smith, Jet Propulsion (approx. 1944)

Neville Duke, Sound Barrier (1953)

Drawings:

Production Concorde

Olympus 602: production powerplant

Fairfield, CT, March 1976

Appendix 9.2

Supersonic and Hypersonic Flight: Research and Flight Milestones

December 17, 1903   First power-driven flight - Flyer I
Orville and Wilbur Wright - Kitty Hawk, NC

August 27, 1939   First jet-propelled flight - Erich Warsitz
Heinkel He 178; jet engine He S 3B designed by Hans Pabst von Ohain - Marienehe, Germany

October 14, 1947   First supersonic flight - rocket-powered Bell X-1
Charles E. Yeager - Edwards AFB, California

July 1949   First supersonic flight by a jet engine assisted by a rocket engine - Douglas D-558; Westinghouse turbojet J34, Reaction Motors rocket engine - Gene May

July 27, 1949   First jetliner flight (subsonic)
De Havilland Comet - England

May 1952   First scheduled commercial jetliner service
De Havilland Comet - London-Rome-Cairo-Entebbe-Johannesburg

December 13, 1953   First high-Mach number flight (Mach > 2) - Bell X-1
Charles E. Yeager - Edwards AFB, California

February 29, 1964   President Lyndon B. Johnson reveals: Several Lockheed jet aircraft A-11 flew at sustained speeds of more than Mach 3 (more than 2000 mph)

May 1, 1965   Official speed record 2070.1 mph - Lockheed YF12A
Robert L. Stephens, Daniel André - Edwards AFB, CA

185

June 5, 1969	First supersonic flight of Russian SST Tupolev Tu-144
September 1, 1974	Transatlantic crossing New York-London in less than two hours (1 hr 55 min 42 sec) - Lockheed SR-71 James V. Sullivan, Noel F. Widdifield
December 26, 1975	First commercial SST service - Tupolev Tu-144 Moscow-Alma Ata, Soviet Union (Aeroflot)
January 21, 1976	First scheduled passenger SST service - Concorde London-Bahrain (British Airways) Paris-Dakar-Rio (Air France)

When it is not necessary to change
It is necessary not to change.

Falkland's Principle

Chapter 10

AVCO Lycoming

A Rotary Engine

The Lycoming Stratford Division of the AVCO Corporation hired me in 1967 for my experience in gas turbines and also in rotary engines. John Marshall had made his name with rotary compressors in England. He also had invented a derivative thereof called the Tri-Dyne engine. Lycoming had acquired two of these engines, giving it the code-name Roto-Lobe. With it Lycoming had hopes to fill the gap between the aircraft gas turbine and reciprocating engines. It was a true rotary machine with no mechanical speed limit. Some of the Lycoming engineers believed that this feature alone would assure achievement of the target. However, tests with an experimental engine were shockingly disappointing.

As my first assignment, I was to assess the potential of the engine on the basis of the available drawings and hardware. I took one look at the drawings, shuddered inside, and mulled over my options. From my analytical, design and test experience with rotary engines, the chances of the Roto-Lobe were minimal, close to zero. To say that right away might have embarrassed the promoters; they might have frowned on my arrogance, and my career within the Lycoming organization would have been jeopardized. So I chose a more subtle approach, suggesting that I analyze the engine in detail and then report on my findings.

The engine geometry, pictured in Figure 10.1, consists of a power rotor with three lobes and of two auxiliary rotors, each with three cavities. The working chambers are interconnected with communicating pipes. Each rotor rotates about its own axis, making the engine fully balanced without any mechanical speed limit. That is its only positive feature!

I compared the Roto-Lobe cycle and operational geometry with those of a reciprocating and a Wankel engine. The following influencing factors were detrimental to the success of this engine: low volumetric efficiency, large volume and mass flow discontinuities, high leakage rates, and high pumping losses. The gas leakage, investigated at the critical positions of the rotors, alone was sufficient to

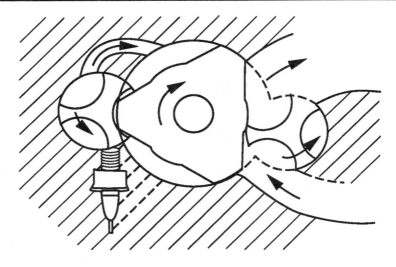

Fig. 10.1. The Roto-Lobe engine geometry.

reject the engine. I had previously established the axiom: If a positive displacement engine exhibits a theoretical leakage path, it can be described with two words — "Forget it." The Roto-Lobe was a typical example to verify this statement.

In a conference with engineering management, directors and specialists, I presented the Roto-Lobe case with colorful charts of pressure-volume diagrams, pressure-time diagrams, the surface/volume ratio of the combustion chamber, the leakage rates resulting from the differential expansion of the rotors and their housings, and theoretical engine performance data. My conclusions and recommendations were simple: Roto-Lobe engine concept feasible, but not practical and/or competitive. Therefore, stop the project. Dr. Franz, Vice President of Engineering and Assistant General Manager of the Division, thanked me for a job well done, and for saving Lycoming further development dollars.

A few fascinating aftermaths followed my report on the Roto-Lobe engine. To this day, some of the engineers present at the conference still tell me I was wrong in discarding the engine. They believe that with enough brain power and money the problems could have been solved. I am adamantly of the opposite opinion. Some problems, I believe, can be shown on the back of an envelope to be insurmountable, and the Roto-Lobe had not only one, but several such problems.

Lo and behold, two years after Lycoming had scuttled the Marshall/Roto-Lobe engine project, the American magazine *Popular Science* published an article with the headline: "Tri-Dyne: Slick New Rotary Engine Could Lick the Wankel." It described a 350 cc test engine delivering 90 horsepower, better than 4 horsepower per cubic inch displacement. It praised the pure rotary motion of the three rotors, which made it lightweight. It also mentioned that two of these engines had been sold to Avco Lycoming in Stratford, and that they are thinking of aircraft applications, but very little had leaked out of the security screen of this company! That was the last I heard about the Marshall/Roto-Lobe/Tri-Dyne engine.

Turboshaft Engines

My subsequent Lycoming activities can be conveniently categorized as follows: solutions to major service problems of the T53 and T55 engine series;[1] layout, design and development of components for advanced gas turbine and alternative engines within the Contributing Engineering program;[2] evaluation of outside proposals and patents; and, last but not least, assistance to proposals in response to government RFQs.

Compressor Disk Failures

In the midst of my rotary engine evaluation, a compressor blade attachment failure in a T53-L-13 engine occurred. I was selected to spearhead the investigation of its cause. A tenon of an aluminum disk holding the blade dovetail root had broken off causing power loss and shutdown of the engine. The single-engine helicopter was thus forced to land by autorotation. Since it happened after a total of 3/4 of a million "trouble-free" flight hours, it was first considered as an isolated case. We soon learned that this assessment is a convenient deception: Every series of failures starts with number one. I clearly remember mowing my lawn on a Saturday morning when I received a telephone call that a second engine had failed in this manner.

We immediately instituted changes and retrofitted engines with these fixes to avert further failures. During this transition period, however, failures occurred at an increasing rate. A Blue Ribbon Committee therefore was established to assess the best solutions to the problem. The committee consisted of representatives of the Army Aviation Systems Command (AVSCOM), the NASA Lewis Research Center, academia, independent institutes, and Lycoming. I was elected chairman of the committee. On the one hand, it sounded strange to me since I was representing the company producing the engines. On the other hand, I felt honored by the confidence placed in my professional integrity.[3]

---

[1] The T53 and T55 turboshaft engine series were the brainchilds of Dr. Franz who had previously become famous with the first mass-produced jet engine, the Junkers Jumo 004.

[2] Research, applied research and component development were traditionally conducted with two programs: Independent Research & Development (IR&D), and Contributing Engineering (CE). The latter became my responsibility shortly after my rotary engine presentation.

[3] Similar "task forces" were to follow; I was chairman of most of them.

Our analysis of the possible failure modes included stress-rupture, high cycle fatigue caused by blade vibrations, and low cycle fatigue (LCF) resulting from the frequent takeoffs and landings of the helicopter. All three items were suspect. The failures starting during Vietnam's hot season indicated that the temperature of the disk rim must have played a critical role.

After meetings at Lycoming Stratford, NASA Cleveland, and AVSCOM, St. Louis, the committee reached the consensus that a major reduction of the tenon stress would solve the problem; it would benefit all three failure modes. I rejected a further suggestion to add mechanical blade damping by fitting the blade root loose in the disk instead of with a press fit, noting: "At idle, a loose blade would provide damping; at operating speed, a weight equivalent to a Volkswagen hanging on it will make a solid attachment without any damping." This statement was effective in shelving the damping suggestion.

Both analytical and experimental results substantiated the safe operation of disks with the lower tenon stress, but only for a limited number of operating hours or LCF cycles. In view of this probability, Lycoming already had designed and certified a drum-type, one-piece welded rotor made from titanium. It was a simpler design and assembly than the original rotor with stacked, mechanically clamped disks.

Cost calculations for the two cases, disk or drum rotor, showed savings for the former. Replacing aluminum disks after a specified operating time with new ones was cheaper than converting to the titanium rotor, and the Army selected this route.

The German defense forces also operated helicopters with this engine. Lycoming had tried to convince them to convert to the titanium rotor for safety reasons. On a mission to Germany, I presented the case for it with technical data supported by a stack of color slides. Though the Germans did not commit themselves immediately, I left with the feeling that I had convinced them. I was wrong! The summer temperatures in Germany being lower than those in Vietnam, they also calculated cost savings by staying with the disk rotor, i.e., replacing aluminum disks with new ones after specified operating hours.

A tragic event taught the proverbial "bean counters" a lesson. An unexpected disk failure caused a helicopter crash in which a high ARVN general, news reporters, and other occupants were killed. The U.S. Army got the message and introduced the long-life titanium rotor.

An official of the German Ministry of Defense, on a U.S. trip, expressed to me in a telephone conversation his frustration that so far he had been unsuccessful with the conversion to the titanium rotor. Cost savings of the engine with the old rotor were too great; the titanium rotor would become cost-effective after sixteen years! I told him, "You do it all wrong with your cost analysis. Use a simpler and more effective method. Invite one of your generals to a helicopter trip over the beautiful

Black Forest. Simulate an engine failure. Don't kill the general; the forced hard landing will scare him enough." The official appreciated my suggestion and promised to think about it. As long as I was in the Lycoming employ, the Germans were still using the aluminum disk rotor exclusively. About two years after my retirement, a Lycoming friend called me. He had a message which would please me; he said: "The Germans had adopted the titanium rotor." I was happy about this outcome. Safety considerations had finally triumphed over dollars and Deutschmarks!

Tank Gas Turbine - Revisited

Since the end of WW II I was actively pursuing vehicular gas turbines culminating in the actual design and development of the Parsons tank gas turbine. This powerplant had demonstrated the advantageous potentialities in a mobile test bed, but also the difficulties for achieving low fuel consumption. Subsequently I went out on a limb by establishing proper perspectives for this technology with my 1952 Mahle treatise. With these in mind, I had monitored the worldwide development in the commercial automotive field and was happy that my cautious views were continually confirmed. In 1965 I was associated at WAD with its proposal for the U.S. Army's 1500-hp tank engine. Naturally, at Lycoming I was extremely eager to have a first-hand look at its AGT 1500 gas turbine, the winner of the Army competition. It caused ambivalent emotions!

High power concentration, expressed as power-to-weight and power-to-volume ratios, necessitates high cycle temperatures and high pressure ratios. Compressors to achieve these higher pressures operated well at the design point but not at lower speed and airflows due to the compressibility phenomena. A variety of methods was conceived and applied to overcome these difficulties: mechanical separation of the low- and high-pressure portion of the compressor; variable inlet guide vanes and variable stators for the front stages; bleed-off; and a combination of these means. The twin-spool turbomachinery was a favored design; Lycoming utilized it in the AGT 1500 tank gas turbine and the same flow machines in its non-regenerative version for helicopter propulsion.

WAD's engine proposal was based on a moderate pressure ratio with simple, rugged turbomachines, and a highly efficient heat exchanger. In contrast, Lycoming had selected a high pressure ratio with high-efficiency turbomachines and a moderately effective recuperator. On the one hand, the aero- and thermodynamics appeared capable of fulfilling all stipulated requirements, particularly that for low specific fuel consumption over the important operating power range. It led, however, to a drastic deviation from the proven configuration of the T53 and T55 engine family resulting in complex, sensitive and costly turbomachinery. I realized that this approach may have been necessary at the time, but I wondered whether it would be destined to completely replace piston engines with gas turbines in the armored vehicle field.

191

The AGT 1500 turbine engine (Figure 10.2) incorporates a two-spool gas generator, the low-pressure (LP) compressor with five (originally four) axial stages, and the high-pressure (HP) compressor with four axial plus one radial stage, each compressor section driven by a single-stage axial turbine. An ingenious cylindrical recuperator preheats the compressed air which obtains its turbine inlet temperature of 2180°F (1193°C) in a single-can combustor. For safe operation and low specific fuel consumption in the full power and speed range, the LP compressor is equipped with variable inlet guide vanes, the HP compressor with bleed-off. For fuel economy at part load and for vehicle braking power, the first of the two axial stages of the free power turbine contains variable stators. Engine power is delivered via a planetary reduction gear to the tank transmission. The HP spool drives an auxiliary gearbox for the engine and tank accessories. This three-shaft arrangement in combination with the concentric recuperator leads to a compact powerplant with the heat exchanger occupying the greater portion of the installation volume.

During the development process, many technological and financial hurdles had to be overcome. My contributions were appreciated.

In 1972, Lycoming's AGT 1500 gas turbine was finally ready to enter the head-on competition with a tank diesel engine for powering the main battle tank XM-1. The Chrysler Corp., the staunchest proponent of gas turbine passenger cars, chose it whereas the General Motors Corp. was conservative and selected a Teledyne Continental diesel engine. After some initial meetings with the Chrysler team on reliability and maintenance items, my major work left me only with the time to watch this fascinating competition, and to wish for a successful result.

Almost simultaneously, Lycoming entered the tank engine without recuperator into the powerplant competition for UTTAS, an advanced military helicopter. Though this PLT-27 turboshaft engine offered impressive installation features, high power and low fuel consumption, the Army awarded the contract to the competition. GE's engine configuration resembled that of Lycoming's T53. The Army apparently preferred this proven technology over the novel three-shaft configuration.

Lycoming's hope for two engines with identical turbomachinery, the AGT 1500 with heat exchange for the tank and the PLT-27 without for the helicopter, was shattered. The danger existed that the Army would have GE add a heat exchanger to its engine for the tank, and thus it would have two engines with the same turbomachines. The advanced development stage of the AGT 1500 eliminated this possibility. GE was apparently able to replace the two-spool gas generator, considered a requisite at the inception of the tank engine, with a single spool, and thus offer a simpler engine. My original concern, though only intuitive, had been confirmed. To the best of my knowledge, apart from the Rolls-Royce RS360 and the AGT 1500, two-spool gas generators are not used in this power class.

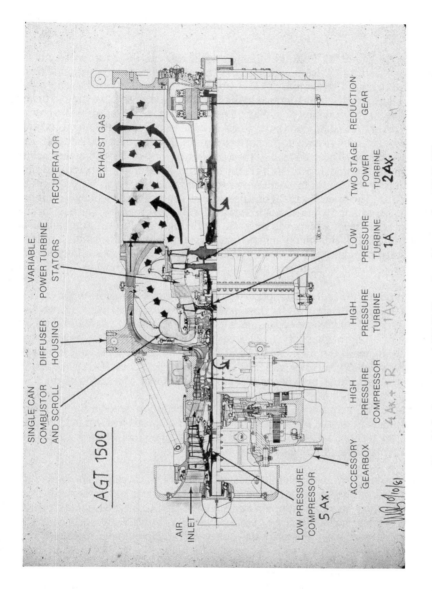

SINGLE CAN
COMBUSTOR
AND SCROLL

VARIABLE
POWER TURBINE
STATORS

DIFFUSER
HOUSING

RECUPERATOR

EXHAUST GAS

AGT 1500

AIR
INLET

LOW PRESSURE
COMPRESSOR
5 Ax.

ACCESSORY
GEARBOX

HIGH
PRESSURE
COMPRESSOR
4 Ax. + 1 R

HIGH
PRESSURE
TURBINE
1 Ax.

LOW
PRESSURE
TURBINE
1 A

TWO STAGE
POWER
TURBINE
2 Ax.

REDUCTION
GEAR

Fig. 10.2. Lycoming AGT 1500 tank turbine engine (1965/79).

193

Automotive Gas Turbine Engines

The ongoing tank engine development also prompted me to review automotive gas turbines in general.

At the 1970 AVCO Pollution Conference I was privileged to represent the Lycoming Stratford Division with a paper "Use of Gas Turbines in Automobiles." As an introduction I stated that our Division had succeeded in selling some 1-1/2 billion dollars worth of gas turbine engines for 2-1/2 reasons: Compared to piston engines they are (1) smaller, (2) lighter, and (2-1/2) free of vibrations. The power/weight ratio of present production engines exceeds that of equivalent reciprocating engines by a factor of two to three; engines in development will double that impressive figure. These advantages are associated, however, with higher acquisition costs on the order of $35-50/hp compared to $17-28 for non-supercharged and approximately $30/hp for turbocharged aircraft engines. Automobile engines sell for $3-5/hp. The specific fuel consumption of gas turbines is likewise higher, especially at part load where automobiles predominantly operate. The best turbine may exhibit in the 20-30% power range 50-100% higher fuel consumption than an Otto engine of equivalent maximum power. The higher cost for acquisition and fuel are partially compensated by lower maintenance costs, greater reliability and durability, and longer life to overhaul.

To overcome the fuel consumption deficiency, heat exchange from the exhaust gases to the compressed air is now preferred over sophisticated cycles with intercooling and multistage expansion with reheat, as was employed in the Ford model 704. Heat exchangers as stationary recuperators or rotary regenerators drastically increase the bulk, weight and cost of the powerplant, and reduce its reliability, durability and overhaul life. Still, the life-cycle costs (LCC) of a mature regenerated gas turbine may become competitive to those of a piston engine.

Examples of automotive gas turbines with rotary regenerators have been demonstrated in the U.S. by Chrysler, GM and Ford, abroad by Leyland,[4] Nissan, and Fiat, among others. Lycoming incorporated a recuperator in its tank gas turbine, so did Daimler-Benz in a truck engine.

Summarizing the investigations I stated: The non-regenerated automotive gas turbine has a promising future in the off-highway vehicle and associated fields. Heat exchange will extend the application to large highway and especially military vehicles. It may happen that city trucks and buses will be equipped with low-exhaust-emission gas turbines if the exhaust of reciprocating engines cannot

---

[4] In the Fall of 1969, British Railroads planned the Leyland gas turbine, fairly advanced in its development, for rail propulsion. Our General Manager wanted Lycoming to negotiate a manufacturing license. When I dragged my feet, he almost fired me fearing we had missed the boat. I kept my cool. Production of the Leyland engine would have been in my opinion premature. In the end, my view prevailed; production never materialized.

satisfactorily be cleaned. The basic disadvantages of small gas turbines for passenger cars, namely high initial and operating costs, lack of facilities for service and maintenance, etc., will most likely not be overcome in the foreseeable future, even under the high political pressure for clean air in the cities.

Low-Power Gas Turbine Engines

The development of gas turbine engine components constituted a great part of my professional career. My work at Lycoming posed new challenges due to the extension of gas turbine applications to the low-power class. I already had been faced with the difficulties associated with the downsizing of gas turbines in general and qualitative terms; at Lycoming they reached a crossroads.

The two-spool, three-shaft configuration was also employed in a Lycoming research project for a 5-pound-per-second engine, the PLT32, concurrently with the Rolls-Royce development of its engine RS360.

These two engine designs, PLT32 and RS360, though independently conceived, exhibit a remarkable resemblance: a multistage axial low-pressure and a single-stage radial high-pressure compressor section, each driven by an axial turbine, and a two-stage free power turbine. This identity was considered as a sign of a definitive advancement of the state-of-the-art for this engine class. The engine complexity and high cost prevented me from sharing this judgment.

Rolls-Royce enjoyed the good fortune that the British Military had earmarked this engine for installation in an advanced helicopter and therefore was paying for its full development and production. We had developed the components to a status that satisfied the specified engine performance. We lacked, however, a specific application and government support. This setback may have been a blessing in disguise. It prompted us to conduct a comprehensive evaluation of gas turbine concepts in the low-power class, under 1000 horsepower. I was happy to have the opportunity to review my previous considerations and express them in quantitative terms.

One significant result of my comprehensive investigations is shown in Figure 10.3. It depicts for typical powerplants the power concentration expressed as specific bulk (hp per cu.ft. of engine installation volume).

The square/cube law holds for geometrically similar or equivalent reciprocating engines, the power increasing with the square of the linear engine dimension and the volume with the cube. Geometrically similar or equivalent positive-displacement engines (Otto, Diesel and Wankel engines) are defined as having the same parameters, number of cylinders, bore/stroke ratio, piston speed, and brake mean

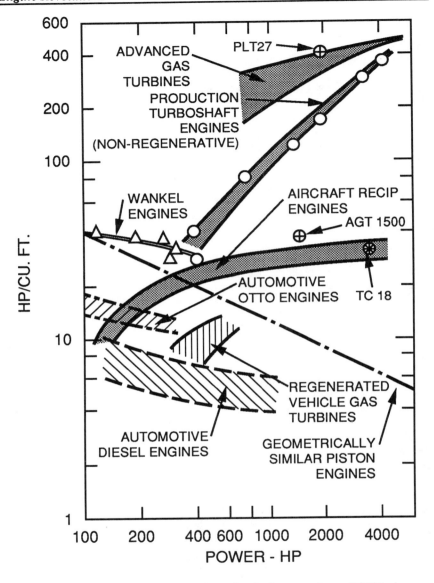

Fig. 10.3. Power concentration of typical powerplants (1970).

effective pressure (BMEP). For gas turbine engines such an accurate, strict definition is not applicable, due to the more complex Brayton working process. This basic difference is explained in my Mahle paper.

Actual automotive Otto, Diesel and Wankel engines follow approximately the square/cube trend; the deviations reflect the differences of the engine parameters. This effect is most pronounced with the aircraft reciprocating engines. With increasing power output, high specific power is achieved by raising the number of cylinders and by increasing BMEP with supercharging and turbocharging as well as by increasing piston speed.

Gas turbine turboshaft engines illustrate a completely different influence of the power output. The specific bulk of a production engine of 4000 hp is with 360 hp/cu.ft. nine times higher than the 40 hp/cu.ft. of a 400-hp engine! At 400 hp, gas turbines are slightly superior to Wankel engines, but penalized with much higher specific fuel consumption.

Regeneration of gas turbines improves their fuel economy but drastically reduces their power concentration. The AGT 1500 tank engine incorporates the same turbomachinery as the PLT 27 turboshaft engine, but its specific bulk is only about one-tenth of PLT 27's 400 hp/cu.ft.! It is approximately on the same level as Curtiss-Wright's Turbo Compound TC 18.

As a result of its high turbine inlet temperature, the AGT 1500 power concentration is still high compared to commercial automotive regenerated gas turbines; these are equivalent to Otto engines and about double that of Diesels.

Fig. 10.3 dramatically demonstrates three items: (1) the superiority of the non-regenerated turboshaft engine in the high-power class, (2) the complexity, volume and cost penalty associated with regeneration, and (3) the formidable competition of reciprocating and Wankel engines in the low-power class.

Aware of these fundamental considerations, we explored all avenues to arrive at a market-oriented gas turbine engine for the low-power class. In our investigations of the scaling of successful high-performance engines to smaller sizes, the variety of detrimental factors were identified and critically assessed, as well as the effects of the physical limits set by the design and manufacturing processes. All items indicated a negative effect of downsizing. Furthermore, development and production costs placed these engines in a precarious position which made them suitable for military applications only.

In view of these results, two teams were established with the assignment of proposing a more competitive, "low-cost" engine in the power class of 5 lb/sec airflow.[5] My colleague Siegfried H. Decher (1912-1980) was in charge of one, I of the other team. Working independently, we arrived at identical engine configurations: a single-spool compressor consisting of one axial and one radial stage driven by a single-stage axial turbine, a reverse-flow annual combustor and a free power turbine. Fig. 10.4 shows the flow path and turbomachines. Lycoming had returned to the simple layout of the highly successful T53 and T55 Lycoming engines.[6]

During our low-power investigations Lycoming received an outside proposal for a low-cost gas turbine based on a differential turbine principle. These configurations were supposed to simplify gas turbine powerplants without sacrificing its major attributes. Like a single-shaft machine, the basic differential gas turbine consists of a compressor and a turbine which are not directly connected with a shaft but with gears and clutches. Claimed advantages were better part-load and idle fuel consumption, faster acceleration, and a broader operating range.

The proposed engine comprised a centrifugal compressor, a single-can combustor, a radial-inflow turbine and a differential gear/clutch arrangement. Low development and manufacturing costs consistent with competitive performance, and versatility to power commercial turboprops, helicopters, vehicles and industrial machines, were claimed. I was selected to evaluate the engine design.

First I was bothered that differential turbines had been around for a long time but had not achieved any application. Second, the stipulated performance depended on three items: Compressor pressure ratio and efficiency, mechanical integrity of the turbine at high gas temperatures and high tip speeds, and the imponderable characteristics of the differential gear and clutch. A detailed assessment of these components led me to the conclusion that the project was highly ambitious; achieving the performance goals was theoretically possible but extremely doubtful with a short and low-cost development period.[7]

Lycoming decided to pursue the differential turbine as a special project with hardware. Various difficulties occurred with the mechanical operation and the performance of the engine. Due to the slow progress it was felt that the expected results could not be achieved within the allotted time frame and budget, and the project was finally abandoned. (This was my first but not my last encounter with a differential turbine.)

---

[5] The term "low-cost" is occasionally applied to expendable, short-life engines; these were excluded from our considerations.

[6] The details of these investigations and their results are documented in the SAE Paper No. 720830 entitled "Evolution of Small Turboshaft Engines" by M. Bentele and Justin Laborde. It was presented at the National Aerospace Engineering and Manufacturing Meeting, San Diego, California, October 2-5, 1972.

[7] My Roto-Lobe evaluation had caused immediate cancellation of the engine development. In the case of the differential turbine, I appeared less convincing about the high risk involved. Furthermore, I was advised to weaken my negative comments.

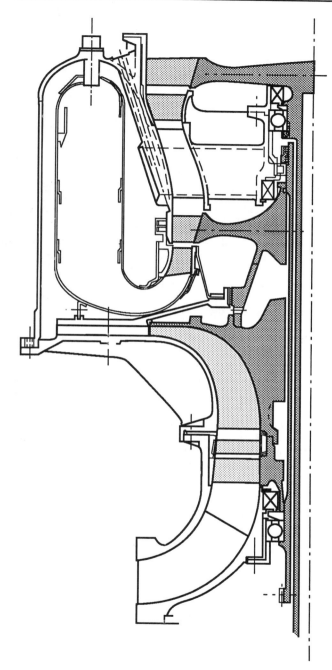

Fig. 10.4. "Low-cost" engine (1971).

As a result of the differential-turbine misadventure, Lycoming pursued the development of the proposed "conventional" engine concept as Model LTS 101 per Figure 10.4. The engine presents an excellent tradeoff between performance, complexity, reliability, and cost. It is produced in turboshaft and turboprop versions with a take-off power range from 550 to 684 horsepower. As of this writing it is operating in the U.S. and abroad in three helicopter and in one fixed-wing aircraft model.

Preparing other proposals for advanced engines from 400 to 4000 shaft horse-power, we often debated and assessed the major design features, single versus twin-spool, axial vs. radial flow compressors, conventional vs. supercritical shafts, and so on, and conducted heated arguments on the pros and cons of individual features. On more than one occasion it came to my mind how fortunate we engineers still are in our field. Within a relatively short time the results of our work are available, exemplified by obtaining (or losing) the engine development contract, or by achieving the engine performance first on the test stand, then in the field, and finally, being gratified by the satisfaction of the users. In contrast, pioneers in economics, social engineering and other faculties are still subject to arguments long after their recommendations are adopted. They may never learn whether their "designs" were successful or not.

Turbine Containment Failure

The July 1989 DC-10 tragedy in Sioux City, Iowa,[8] reminded me of a similar accident that Lycoming had experienced in the Fall of 1970 with a T55 turboshaft engine in a Vertol Chinook helicopter. Blades of the power turbine had broken off; the bearing support had been unable to contain the highly-unbalanced turbine wheel which then severely damaged the control system of the helicopter.

The bearing support housing had to be strengthened to eliminate these secondary failures. At the time, a combination of design, stress analysis and full-scale testing was the sole means to arrive at a satisfactory solution, and to verify it. Under my jurisdiction of the integrity testing, the failure mode was simulated for the worst case, the maximum power turbine operational speed. A power turbine blade root was weakened to such a degree that it would fail at this speed. In view of the great expense of these tests, complete loss of the power section and other damage to the engine and test stand, I devised an iteration method with which stress-rupture blade failures were induced at the exact speed, thus reducing the number of necessary tests to a minimum.

---

[8] Fatigue failure of a fan disk had caused severe damage to the hydraulic systems resulting in almost complete loss of aircraft control. The crew heroically managed to keep the aircraft in the air to reach the nearest airport. During landing a wing sheared off and caused a fire. Miraculously, many passengers and crew survived the ordeal.

To the best of my knowledge, this failure test program prevented Lycoming from experiencing a reoccurrence of containment failures. The T55 became the core of the high-bypass turbofan ALF 502 that powers the four-engine quiet aircraft BAe 146 of British Aerospace.

Retirement from Industry

Lycoming's labor relations policy decreed mandatory retirement at the end of the month following the employee's 65th birthday. It was the custom that the colleagues bid farewell with a retirement party.

When my retirement date drew nearer and nearer, I was unable to notice any signs for such an event. My 65th birthday also was ignored. My best friends chose to evade talking about the forthcoming separation from Lycoming. I expressed my disappointment to my wife; she tried to cheer me up.

Shortly before my departure date, Lycoming had a visitor from Germany whom I had known during my Stuttgart days. I participated in the discussions, was invited to have dinner with him, and to bring my wife. We drove to a country club, a friend opened the door — and wow! Over 100 people cheered us. They represented a cross-section of Lycoming employees, from the mahogany offices to the shop floors, most accompanied by their spouses. They had gathered to surprise me with a testimonial dinner in my honor, with an elaborate printed program. My friend Decher, diligent organizer of the event, had sworn to secrecy all his volunteers, including my wife. The party highlights were: the presence and speeches of three guests from Curtiss-Wright, the Champion Spark Plug Co., and Daimler-Benz, Stuttgart, whose cooperation I had enjoyed in Germany and in the U.S.; a book with congratulatory and best wishes telegrams and letters from the U.S., UK, Japan and Germany; talks, speeches, and skits by my superiors and peers of my most famous engineering works and problem-solving endeavors, i.e., the life-saving Jumo 004 blade vibrations calamity, the Marshall/Roto-Lobe engine development and me stopping it and similar ones with a sickle (Figure 10.5), the T53 disk failures, and my detection of mistakes in official documents. Color slides, program documentations with huge-size charts, and two songs by the engineering mixed chorus enlivened the proceedings.

A 2-1/2x3-1/2-foot retirement card, painted by the Lycoming house artist, included my portrait, the citation "Engineer, Inventor, Educator, Editor, Sports Enthusiast," and paraphernalia (tennis racket, skis, fishing rod, garden plants and a bible with a note "Matthew 5:3"), was signed by my friends, 250 in all. This card initiated the presentation of farewell gifts. Then followed bookends made with the time-limited aluminum compressor disks, and for my mathematical mind two

Fig. 10.5. One of the skits presented at the Testimonial Dinner (1974).

Fig. 10.6. Magda and me, thanks and farewell.

calculators: first a 2x3-foot abacus handcrafted by a friend with fine wood and 22 white and 55 yellow tennis balls, washed and tumbled by his wife; and second, a novel electronic desk calculator which was hidden in the wrapping of an old tennis racket.

It was a great farewell party. Everybody enjoyed it, particularly my wife and me (see Figure 10.6). I thanked each and every one, also on behalf of Magda, for their efforts and love, with tears in my eyes. Such a surprise party was the first and only one ever conducted at Lycoming. My colleagues would not dare to do it with anyone but me!

Chapter 11

Post-Retirement Pursuits

Commercial Ventures

After some forty years in industry employment, I planned my retirement with the motto noted above. Most of my friends agreed with me as expressed in their telegrams sent for my retirement party: You're not likely to retire; starting a new career; productive retirement; you'll not stop contributing; survival at Curtiss-Wright has proven your capability of surviving anything including retirement. Also, companies and institutions appreciated my experience in specific fields, and wanted to use my services and pay for them.

My contributions to the technology of jet, gas turbine and rotary engines had failed to make me independently wealthy. The CW, NSU and Wankel companies collected tens of millions of dollars in license fees and royalties. Berner and von Heydekampf praised my invaluable efforts to the Wankel engine success but were unable or unwilling to let me participate in the incoming windfall profits. At CW the distribution of a portion of the economic benefits and/or royalties to employee inventors had been discontinued; it was replaced by nominal Patent Incentive payments. Berner promised to equal the amount I would receive from NSU and Wankel, the two prime licensors. In Germany employees are by law entitled to appropriate compensation for their patents assigned to their employer. Because I was neither an employee of NSU nor Wankel, von Heydekampf felt no obligation to pay me. In short, both Berner and von Heydekampf were hiding behind alleged legal barriers.[1]

---

[1] This area is also discussed by independent observers. In his book <u>Memories and Machines: The Pattern of My Life,</u> Sir Harry Ricardo describes the diesel engine as the justly claimed, most efficient form of prime-movers. He continues: "Since my schooldays controversy has raged as to the proportion of credit due to Diesel and Ackroyd-Stuart for its initiation. My own view is that the lion's share should go to the very able group of engineers at the German firm of M.A.N., for it was they who made the engine work in a practical form." Another Englishman, Andrew Nahum of the London Science Museum, Aero Engines and Propulsion, wrote in an article in the magazine *Fast Lane*, January 1986, on the Wankel engine development: "Apart from Dr. Felix Wankel with his original 'double-rotary engine,' there were three 'nursemaids' who were critical to the Wankel's progress, Dr. Walter Froede of NSU, Dr. Max Bentele of Curtiss-Wright and Dr. Kenichi Yamamoto of MAZDA."

The CW retirement plan did not contain a cost of living adjustment (COLA) provision; the two-digit rate of inflation had caused a drastic decrease in actual worth of my monthly benefit check. This erosion continues at present at a slower rate.[2] Also, I was allegedly a few months too old to be eligible for the generous AVCO retirement/health insurance package.

Major items of concern presented the full-of-pitfalls, vague regulations about my earned income, the hours per month and the hourly rate, the associated mandatory contributions to and the benefits from the Social Security Administration, federal and state taxes, the provisions for health insurance, pension plans, etc. This tangled, murky situation prompted us to organize the business activities properly; my wife and I founded our own company and incorporated it for engineering research, consulting and writing. With parts of our given names, Max backwards and Magda forwards, we called it XAMAG, Inc.

Our first official business took place exactly a fortnight after my departure from Lycoming. My wife and I were close friends of two junior members of the Braun family of Wolfstein in the German Palatinate. Hans Braun (1899-1987), the patriarch and his family owned and operated a factory that was founded in 1903 by his father Karl Otto Braun (1862-1921), with his name on the building. Hans had led the enterprise since 1921 through peace and war to its supreme position, and he had received many honors and awards from the local, state and federal governments as well as from civic and trade associations. The company produces special fabrics for medical and technical purposes, and constitutes one of the most important enterprises of this kind in the world, with exports to 48 countries, including the U.S. To be able to better serve that market, the Braun family considered the acquisition of a small American bandage factory. I was asked to offer my experience in American industry and assist in the search. Finally, the desirable plant was found in Bridgeport, Connecticut, a few miles from our Fairfield home. The Brauns then elected me Vice Chairman of the Board. As such, I participated in the signing of a stack of documents sealing the purchase and related agreements of the plant, renamed CONCO Medical Company. The date was Valentine's Day, 1974.

Textiles were, of course, a new field for me. As engineer, I was sometimes obliged to use the expediency of fixing engine problems with "Band-Aids." So getting to know the real thing was an exciting challenge.

To modernize the CONCO manufacturing facilities, specialists and technicians from Germany assisted in this conversion. We familiarized them with the American

---

[2] In November 1983, CW terminated its Post-Retirement Health Care Coverage for former WAD employees only. Court action initiated by a formed Retiree Organization against this discriminatory action still lingers on as a class action suit. As of this writing, a trial and judgment are out of sight. The seven-year delay is caused by the judicial system being overloaded with criminal cases that receive priority treatment as a matter of law. For our protection, we pay in the meantime for our own health insurance.

way of life and mores, and helped them to cope with tricky situations such as arson of the guest house, burglaries, shootings, etc. Speaking their language and offering a shoulder to cry on made them feel at home; they appreciated our assistance and concern immensely.

Special occasions presented visits of the Braun senior and junior family members who deserved red-carpet treatment. Business and pleasure was then combined, inspections of the plant and discussions with its staff and workers, board meetings in our home, visits to subcontractors and then dinners, automobile tours into the country and shopping in the nearby "Big Apple," New York City. On one occasion, three-and-a-half generations of the Braun family enjoyed their stay in Connecticut and the U.S.

After some early downs, red numbers, and a new top management, the future looked promising. The plant became too small necessitating one and then another building expansion. Contacts and negotiations with architects, contractors, and city and state officials provided new experience and knowledge on how governments work — or don't.

The CONCO products may not look high-tech, but the manufacturing processes certainly have to be for this highly competitive market. Our activities contributed to the growth of CONCO which established itself as "The Bandage Specialists." The Braun family, Partners and CONCO honored me, Magda and our family with a reception on my 80th birthday where I was presented with a silver plate inscribed, "In grateful recognition of friendship and services."

CONCO was not my only post-retirement venture. The Mahle company of Stuttgart, the pioneer of aluminum pistons for internal combustion engines, was the leading independent manufacturer in Europe and enjoyed exports worldwide. At the request of its American customers, Mahle considered building a piston factory in the U.S. In view of my long association and friendship with Dr. Ernst Mahle and with some of his associates, my CONCO experience and my general knowledge of American engine industry, I was elected President of the new enterprise, named Mahle, Inc., in 1976. I served mainly as Mahle's legal representative during the search for a location, the construction of the plant, and its initial operation. The Honorable Ray Blanton, Governor of Tennessee, and his Department for Industrial Development had played a major role in the final selection of Morristown, Tennessee, for the factory. The officials of Morristown City and Hamblen County, Industrial Boards, Chamber of Commerce, the Congressman of the district, among others had conducted the spade work; all were present at the official announcement in December 1976, rejoicing this unique, large addition to the industrial economy and expressing the pledge to work for a smooth and speedy construction and peak efficiency operation. Bold newspaper headlines and special Mahle sections reflected these emotions of happiness. My Mahle colleagues and I relished this warm welcome.

The governor formed a Tennessee Industry Transition Task Force to smooth the way for international companies locating in his state, for dealings with the state government, for providing the factory access roads and utilities, and for hiring and training of its workers. Mahle was selected as "guinea pig" for this operation. I enjoyed my participation in this official announcement, its media coverage, and the following reception in the elegant governor's mansion. It led to personal contacts with the governor and his aides; they became the highest government officials with whom I was on a first-name basis.

After working with the architect, prime and subcontractors and solving a myriad of problems, we were relieved when the first pistons came off the machining line in April 1978.[3]

A highlight was the official plant opening and dedication ceremony in September 1978. It was attended by the Governor, state and local government officials, the press and people who had helped us in different ways. The U.S. Congressman for the district presented a U.S. flag which had flown over the Capitol in Washington, D.C. I accepted this honorary symbol saying that Mahle Morristown has become an integral part of this community, Morristown, the state of Tennessee, and the United States.

My horizon was extended in the purely commercial field; being fiercely competitive, it requires application of high technology to the same degree as in the glamour industries. In both cases, the "corporate culture" is different from that of the military/industrial complex. Cooperation between all organizational levels is admirable and a joy to participate in.

Both enterprises, CONCO and Mahle, flourish and grow. I am happy and proud to have assisted in their entry into the U.S. and their continued contribution to the health of its economy.

On the occasion of the German celebration of "50 Years of Jet-Powered Flight," the Minister for Economics, Middle-Class and Technology awarded me in a ceremony in Stuttgart the *Wirtschaftsmedaille* (Medal of Economics) for outstanding merits to the economy of Baden-Württemberg. These services included my work at Hirth and Heinkel in establishing high-technology engineering of turbochargers and jet engines, realistic assessment of vehicular gas turbine and alternative engines, and valuable assistance of German companies in setting up factories in the U.S. beneficial to the economy of both countries. Colleagues, friends and my wife were invited to this presentation. The Lord Mayors of Stuttgart and Ulm, among others, congratulated me for this high honor.

---

[3] One of them adorns my office. I once showed it to defense industry friends, asking them for an estimate of its cost. All missed it by large factors. When I pointed out this discrepancy, they rebutted, "We could not produce at your low cost but we could sell it at our high price."

The Wankel Engine - Postscript

In the mid-1960s, the U.S. was ready to take the Wankel engine seriously. Curtiss-Wright demonstrated a test car with a Wankel engine, NSU's family sedan Ro 80 with a twin-rotor engine became 1968 "Auto of the Year." It suffered initially from difficulties with the apex gas seals which were finally overcome. The success of the rotary engine development team, headed by Dr. Kenichi Yamamoto, enabled Toyo Kogyo of Hiroshima, Japan, to field test the rotary-engined sports car Mazda Cosmo, and then to market it in 1967.

"Wankel fever hits Detroit," blared a magazine's headline. Wankel-equipped cars rolled off the assembly lines in Europe and in Japan. It was just a question of time until American automakers joined the rotary bandwagon.

That event occurred in November 1970: The General Motors Corporation took the 21st NSU/Wankel license for all applications, except aircraft engines. The price was a staggering $50,000,000 to be paid in defined annual installments. GM had the option to quit in case the engine was, in its opinion, unsatisfactory for its purposes.

At the following 1971 SAE Congress, a few friends asked me whether I would move to Detroit to lend GM a hand. That was pure speculation on their part. I imagined that GM felt capable to develop the engine on its own and make its own mistakes, without any outside help. And so it was! Leisurely, I kept watching the Detroit progress mainly by the covers and headlines of American magazines (see Appendix 11.1).

Ward's Communications, Inc., Detroit, inaugurated in July 1972 a biweekly eight-page magazine, *Ward's Wankel Report (WWR) - Assessing the Impact of the Rotary Engine Revolution.* Its last page regularly introduced "Rotary Men in the News." My turn in the sequence was headlined "Max Bentele's Historic Role" and presented my professional life in a nutshell.

The renamed Mazda Motor Corporation introduced other sports and family cars, all with twin-rotor engines. Production rose steadily; the 500,000th rotary-engined car left the Hiroshima factory in June 1973. No wonder von Heydekampf called Toyo Kogyo a gold mine; CW enjoyed raking in license fees and royalties, too.

After a while of GM watching, I discerned an atmosphere similar to the one that I had experienced at Curtiss-Wright a dozen years before, and had then characterized as *Tores-Schluss-Panik*. In the Fall of 1974, GM announced stoppage of the Wankel project with the statement: "The anticipated stringent regulations on exhaust emissions and fuel economy had made it doubtful that the Wankel engine could fulfill these standards."

The GM decision was the bellwether for practically all other American industries to shy away from Wankel engine developments. "If GM thinks it can't be done, why should I try?" the CEO of a marine engine company told me.

The 1973 oil embargo and resulting fuel crisis in the U.S. had caused a setback for Mazda, but its continuous production of rotary-engined passenger, sports and race cars represents the shining symbol of Wankel's invention.

Investigations of a heavy-fuel Wankel engine with the classic Diesel principle were continued in Germany, Great Britain and Japan. Single-stage and two-stage configurations were built and tested, in sizes up to 1200 cu. in. displacement. Results proved unsatisfactory and finally forced the abandonment of all of these programs.

In contrast, my stratified-charge concept proved viable for burning a variety of fuels from gasoline to diesel oil and jet fuels, and attracted both military and commercial interest. At Curtiss-Wright, some difficulties occurred for some applications in the attempt to optimize the process for the full operating ranges of power and speed. In 1976 they were overcome by adding a pilot nozzle to the fuel injection system. DISC multifuel stratified-charge engines were then developed in two- and four-rotor configurations delivering 750 and 1500 horsepower, respectively.

The British, particularly their Fighting Vehicle Research & Design Establishment (FVRDE), also showed interest in this multifuel engine type. This news elated me, although I felt some ambivalent nostalgia. In the late 1940s and early 1950s I had cooperated with FVRDE in the development of the first tank gas turbine. Now they encountered a second novel engine, one which I had conceived and pioneered.

Jet fuel operation proved also important for aircraft propulsion. Curtiss-Wright and NASA conducted relevant studies and tests with experimental rotary DISC engines.

In early 1984 the John Deere Company acquired all North American Wankel rights from Curtiss-Wright. Experiments were conducted with multifuel stratified-charge engines with the new designation SCORE (Stratified Charge Omnivorous Rotary Engines). For example, a 400-hp twin-rotor turbocharged stratified-charge engine designed for aircraft installations met or exceeded all requirements on the test stand. Unfortunately, as a result of the catastrophic decline of the General Aviation market, production plans had to be temporarily postponed.

SAE selected Charles Jones, my successor at Curtiss-Wright, as co-recipient of the 1987 Edward N. Cole Award for his contribution to the stratified-charge rotary engine. The other award recipient was Yamamoto, then President of the Mazda Motor Corp., for his outstanding contribution to the rotary engine technology.

Yamamoto had spearheaded Mazda's development to successful mass-production of rotary-engined passenger cars, had written a number of articles in the professional press as well as the book <u>Rotary Engines</u> in 1971, and updated it in 1981.

Mazda's automotive Wankel engines demonstrate high power concentrations. Their 1990 production engines are offered with specific outputs in hp/cu.in. which are in the naturally aspirated version 40%, and in the turbocharged version 25% higher than the best reciprocating engines.

For other commercial and military applications, the stratified-charge Wankel engine will be the only one capable of penetrating and surviving in these markets in the long run. I am happy and proud to have laid the foundation for this development. An enthusiastic promoter would accelerate it.

One detour — air-cooling of Wankel aircraft engines — still bothers me. In view of its rotor housing "hot spot," that cooling method is really unsuitable for high performance. However, the airplane companies had rejected liquid-cooling. "We would step back 50 years," they said. Curtiss-Wright reluctantly accepted their position and developed air-cooled gasoline and DISC engines. They constituted engineering marvels, but at the expense of cost and sensitivity. Fortunately, better basic thinking finally prevailed and led to the DISC engines mentioned above. Liquid-cooling is again successfully applied to aircraft reciprocating engines; it contributed to the triumph of the nonstop flight around the world of the Voyager in 1986.

Rotary Engine Potpourri

My fundamental work on the Wankel rotary engine, which I personally reported at numerous SAE conferences and section sessions, in talks to societies and clubs, and wrote about in articles in the professional, national, and foreign press, contributed to my reputation, as well as to my nicknames "Mr. Rotary," and "father" or "grandfather" of the American Wankel engine.

The advent of the Wankel engine in the early 1960s triggered a flood of other rotary engine inventions. Since that time I was privileged, or saddled, with the labor of assessing many of these during my employ at Curtiss-Wright, Avco Lycoming, and afterward as consultant to WWR and to companies in the engine field. My adjunct library contains scores of them in the form of press announcements, descriptions, patents, and requests for evaluation.

In early 1974 I wrote in WWR: "Scores of new engines have been and are being suggested continually. Most of them do not get beyond the drawing board or patent office, some reach the hardware stage, then falter on the test stand, some spend time and money in development, but only one has penetrated the market and finds it tough going there (the Wankel)."

My collection at the National Automotive History Collection of the Detroit Public Library contains details and my assessment of some of them. It would fill pages and pages of this book and also bore the reader to deal with all of these engines. Therefore, I will touch upon only a few famous or infamous ones with which I was personally involved. (My negative assessment of the Marshall Tri-Dyne (Lycoming's Roto-Lobe) engine is already discussed above, and also its subsequent revival in magazine articles in the late 1960s.)

At a government-sponsored symposium on alternative automotive power systems (AAPS) in 1973, the head of the Swedish delegation presented a rotary engine under active development at the Svenska Rotor Maskiner AB (Swedish Rotary Machine Co.). It was derived from a two-rotor compressor, with labyrinth instead of rubbing contact seals. With this deficiency in mind, I expressed my serious doubts on the alleged performance, weight and size. But I was told the test results will speak for themselves. Later on, the company issued a colorful brochure and picture of this R-Engine, adorned with the Swedish flag, and predicted that the engine would be a great success. On behalf of WWR, I evaluated it on the basis of its geometry and working principle (Figure 11.1).

As in the case of the Marshall engine and other concepts derived from rotary compressors, I found that the effects of internal combustion are usually detrimental and thus negate the advantages offered by a rotary compressor. SRM's R-engine does not appear to be an exception to this general rule; its pros are moderate, its cons quite substantial; the aerothermodynamic and leakage losses, and combustion will prevent it from becoming a great engine in this era of fuel economy and ecology.

On the other hand, it is immaterial how I evaluate the engine's potential (although my track record in this respect is pretty good), because one day SRM's rotary engine will be on exhibit in Stockholm, Hanover, Detroit, Tokyo, Moscow or Peking. Its specification folder will contain numbers for horsepower output, with altitude and temperature corrections, specific fuel consumption over the full operating range, including idle, installation weight and dimensions, exhaust emissions, maintenance in man-hours and parts costs per engine operating-hour, reliability, times between malfunctions and suggested overhauls, overall service life — and a price tag supplemented by appropriate discount rates for large-volume customers. A stunning blonde waving the Swedish flag will explain all the goodies of the engine, and then the question will be answered — whether the visitors just stand there to watch the girl, or are willing to place orders on SRM's R-engines.

Twenty years after I had evaluated the Marshall Tri-Dyne rotary engine and rejected it on the basis of my experience with the Wankel engine, I received a press report of an interview which Dr. Felix Wankel himself had given to the *Auto Zeitung*, with the headline: "Wankel II: 400 PS aus dem Schuhkarton" (400 horsepower from a shoe box). If I had not seen it with my own eyes, I would have ignored

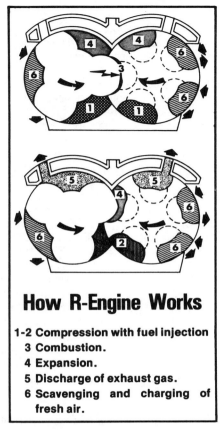

How R-Engine Works

1-2 **Compression with fuel injection**
3 **Combustion.**
4 **Expansion.**
5 **Discharge of exhaust gas.**
6 **Scavenging and charging of fresh air.**

Fig. 11.1. Swedish R-Engine (1974).

it as a bad dream. The Wankel II engine was in principle identical to SRM's R-engine. Initially, that engine's male rotor had four lobes and the female eight cavities. The final version employed three and five, respectively. Wankel had gone one step further and reduced it to two and four. It was debatable whether this change would ameliorate the Wankel II enough to assure its success. My close analysis convinced me that it would not. Wankel himself had now violated most of the principles he had laid down for the practicality of a rotary engine, and had thus tarnished his reputation as a great inventor and "Mr. Gas Sealer." Maybe he thought that at the proposed high speed of 40,000 rpm seals were unnecessary. Since I was *persona non grata* to him, I was unable to engage him in discussing his last brainchild. In his favor, I can only assume that the Wankel II, which presented in

213

his own word his legacy to the motorists of tomorrow, was conceived under the detrimental effects of a kind of Alzheimer's disease. We will never know; Wankel died a year after his announcement of the second rotary engine revolution, and this engine with him. It has not been heard of since.

There actually already had been a second generation of rotary engines. In the Summer of 1972 the Karol engine was announced with a ready-for-production prototype as a first "disposable" pollution-free car engine. Its configuration and operation were kept secret, as well as its patents, which "would only have confused the experts." The engine was under joint development in the U.S. and in Great Britain. I was asked to participate in view of my rotary engine expertise. Despite my expressed willingness to render my services, I did not hear about the engine until *TIME* magazine published an article on December 1, 1972, with the headline "Rotary with a Twist" and a diagram entitled "Split Cycle Rotary Engine." Two items irritated me: first, that I was never called to look at the engine, and second, that the engine was clearly incapable of producing the claimed characteristics. I expressed my displeasure in a letter to the Editor of *TIME*, which WWR printed within my article on the engine. I wrote: "Dear Sir: After the secrecy wraps have finally been taken off the Karol engine, I found only one thing wrong with it, namely your placing it in the SCIENCE category, whereas it should be under SHOW BUSINESS." I also equated the split cycle with a splitting headache. A number of my friends took me to task for being so publicly harsh. In the Summer of 1989 I was glad that the "Cold Fusion" experiments received a similar treatment in very prestigious magazines. The unconventional announcement and the absence of reliable data for abundant energy from cold fusion had prompted the statements that the affair was more entertainment than science. I felt exonerated. The jury on the cold fusion is still out. The Karol principle did not make it as an engine nor as a compressor; it was never heard of again.

My evaluation of dozens and dozens of rotary and other engine proposals provided me with a perspective of their inventors. They can be essentially divided into three categories. The first includes educated or self-made engineers who want to become rich and famous with their inventions. They succeed fairly often in acquiring a sponsor to develop their ideas, in government agencies, universities, or through a rich patron. I find it most difficult to convince them that the laws of the physical sciences cannot be violated with impunity, and that the marketplace provides a most hostile environment.

Typical for this category were the Marshall engine, SRM's R engine, NAHBE (Naval Academy Heat Balanced Engine), and the K-Cycle Engine. The K-Cycle Engine Company in Winnebeg, Manitoba, developed its engine for years with money from stockholders and local and state government agencies. On order of an American venture capital company, I had evaluated the concept and found it theoretically feasible but in practice saddled with almost insurmountable problems.

Even then, the benefits would be small. My detailed technical report was ignored, and after a few more years of "development," the project ended in bankruptcy of the company.

The second category includes rich people who have inherited or made a fortune and want to complement their status with an invention named after them.

The third category are the tinkerers, who pursue their invention as a hobby, but still with the hope they will become rich or famous or both. My evaluation of these novel engines and my relationships with their inventors provided me with experience to put them in the proper category. I shall leave it to the reader to do the same. I am reluctant to disclose my selection except for one typical case.

An inventor in Chile was not convinced about the analysis and evaluation I had submitted to him in writing. He called me on the phone asking for a personal meeting so that he could sort out the differences between his and my point of view on his novel engine. I agreed to receive him in my office in Fairfield, Connecticut. Upon his arrival after his long flight from Chile, I tried to explore his background, particularly whether he worked for a company there. He replied: "No. Together with my brother I own a mine, and I operate it." Thinking of the expense he had already incurred with his invention, and the additional costs if he wants to pursue it further, I said: "It had better be a gold mine." He replied: "It is." He clearly belongs in the second category.

To present the reader with an idea about the puzzling preoccupation of a portion of mankind who wants to invent a novel engine, I have added all engine proposals I had rudimentarily evaluated to the list in Chapter 12, Appendix 12.2.

Automotive Engines for Ecology and Fuel Economy

Automotive engine technology presents a fascinating subject. Three power-plants almost exclusively dominate the marketplace: the positive-displacement internal combustion engine in its Otto or Diesel version, the gas turbine/jet engine, and the rocket engine. Gas turbines and rockets reign unchallenged in air and space, except for small general-aviation aircraft. Efforts to dislodge the conventional engines on the ground and on the sea, continually conducted in the past, have again received emphasis in recent years.

These searches for a powerplant superior to the present ones have failed so far. The previous chapters deal with some of these developments and pinpoint specific reasons that eluded their permanent application. Here I will investigate their difficulties in fundamental terms.

Scientists and engineers have categorized engines that generate mechanical power by burning fossil fuels in various ways, with emphasis on the thermodynamics or the combustion. Examples are: E engines and Q engines; one working medium, ambient air or a second one, steam, vapor or a gas (helium or hydrogen); working cycle intermittent or continuous; combustion internal or external, and so on.

I have occasionally added my own classification to this debate. In all heat engines, heat energy is transferred to the working medium. In true internal combustion engines, the air is heated directly. Heating of the engine parts is kept to a minimum, exemplified by Rudolf Diesel's first concept of a fuel-efficient engine and the recently emerging low-heat-rejection (LHR) engine. In contrast, in the other powerplants (the external combustion and regenerated gas turbine engines) heat energy is transferred through a wall. The heater material places an upper limit on the cycle temperature thereby restricting the attainable thermal efficiency. Furthermore, the vehicle's fuel economy suffers due to the heat that is stored in the heater and the regenerator and is not recovered.

A further difference provides the effect of the power output on the working cycle. The ideal cycle may stay independent of the load or it may quantitatively change. In the latter case, the thermal efficiencies will be lower for off-design operating conditions. The gas turbine presents a typical example for this characteristic as previously discussed.

Toward the end of the 1960s, the national debate on the environment finally led to the Clean Air Act of 1970. It presented the U.S. government with a dilemma. The law of the land was clear, but the technology to fulfill it was missing. It was felt that the automobile industries, which were not engaging in serious work on alternate engines but had "sweetheart contracts" with the internal combustion engine, needed a stimulus. The established Environmental Protection Agency (EPA) therefore instituted a program on Alternative Automotive Power Systems (AAPS), with the goal of demonstrating clean powerplants in a car in 1975. The North Atlantic Treaty Organization (NATO), already having a Committee on the Challenges of Modern Society (CCMS), was induced to participate in this Low Pollution Power Systems Development program (LPPSD). Originally the whole range of nonconventional engines was considered, but subsequently narrowed to the Brayton cycle (gas turbines) and Rankine cycle (steam engines). Then the program was widened to include studies on alternative fuels, electric cars, and miscellaneous other engines.

President Nixon's energy independence program which President Carter continued with MEOW (Moral Equivalent Of War) caused redirection of these government programs; their sponsor then changed to ERDA (Energy Research & Development Administration) and DOE (Department of Energy). The goals for alternative engines were always identical but very vague, a 30-40% increase in fuel mileage over that of a future conventional production car (for which finite numbers were, of course, unavailable) and exhaust emissions lower than future Federal standards. In view of these predicaments, the former lofty meeting title was changed to ATD (Automotive Technology Development).

As a contributor to *Ward's Wankel Report* (WWR), and as a consultant to companies in the automotive field, I monitored these programs, attended the semiannual or annual Contractors Coordination Meetings (CCM) of the American and foreign contractors, and participated in the discussion and evaluation of the engine concepts, experiments, and progress.[4] My position was unique. For the second time in my career (I had enjoyed the first one during the interregnum after WW II), I was free to express my mind, without the strings of corporate policy and the whims of a superior.

In this respect, I found another forum for the uninhibited interchange of ideas. The Champion Spark Plug Company of Toledo, Ohio, conducts two-day meetings entitled "Ignition and Engine Performance Conference" to which all engineers active in the automotive engine field are invited — including its competitors. The conferences are held in the U.S., Canada, Mexico and Europe. I had originally presented portions of the following descriptions of alternative engine developments at those conferences.

I participated in my first conference in London in 1974, and chaired one section in Brussels in 1984. My concluding remarks, made at the Mexico City conference, illustrate their atmosphere: "Today I should like to repeat our vote of thanks which I had instituted some years ago, but I am happy to go a step further and include in our appreciation the Mexico City people here — nuestros amigos Mexicanos. They made our stay a pleasant one. As long as we participants were glued to our chairs and to the words of the speakers, and may I add at most ungodly early and late hours, their gracious wives took care of our spouses, showed them the sites of Mexico City, took them to lunch and to shopping and helped them to economize in spending money. Our Mexican friends also offered to help us participants in exploring the city and enjoying its culture, history and surroundings. In conclusion, primero, un voto de gratitud a nuestros amigos Mexicanos (los recordaremos siempre), and second, I propose a thunderous round of applause to the whole Champion organization. These conferences are unique, and they are voluntary. There is no big brother and no press to watch you which is particularly remarkable in this (Orwell) year 1984. The delegates are free to exchange technical information and speak their mind to solve common problems in the search for greater progress. The response of this meeting indicates that they are desirable. The last two days were fruitful due to the good organization conducted by Dave, his colleagues and his staff, ladies and gentlemen alike. And the interpreters did a good job, too. On behalf of all delegates and of all chairmen, I thank the Champion organization for the invitation to the conference, for the assistance and hospitality extended to all of us. Let's confirm that 'thank you' with a round of applause." Dave's answer: "Max, as I have said at previous occasions, we could not have held this conference without the support and participation of all you delegates. Our thanks go to you!"

---

[4] Initially, attendance, including the "handouts" for the discussed topics, social hour and final report were free. Later on, a fee for noncontractor attendees was introduced, escalating from $25 to $150.

Getting back to the government-sponsored demonstration programs, some appeared farfetched but they served a useful purpose. The ecology and energy problems are so serious that all avenues have to be explored. Doing it on paper alone won't satisfy everybody, as can be gleaned from previous chapters. Also, the cost of such demonstrations appeared minimal in comparison to the total Federal budget and could, in view of the above fundamental considerations, be considered as welfare programs for the participating contractors.

Concern about air pollution, the ozone depletion and the greenhouse effect will continue to influence all mobile and stationary power generation to an increasing degree in the future.

Alternate fuels such as methanol, gasohol and natural gas are suggested as a first step. From 1974 to 1976 I participated at the Stevens Institute of Technology in a government-sponsored study on the use of hydrogen in transportation engines. We found that the difficulties occur not with the powerplants themselves but with production, storage and safety of hydrogen. The Hindenburg syndrome still has to be overcome. I have inspected and driven some hydrogen-fueled vehicles but I feel that as long as the on-board hydrogen storage problem eludes satisfactory solutions, hydrogen will remain exotic for the foreseeable future.

In summary, future powerplant developments will be determined less by the engineering profession, and more by the representatives of the general public in the governments of the world. It will be fascinating to watch the political, economic and financial ramifications of this quest for a clean environment and atmosphere.

Steam Engines

During my college years in the Depression, I was lucky to have a summer job as a fireman on a steam locomotive. It was good money, but hard work. Though not yet fully familiar with the intricacies and ramifications of the laws of thermodynamics at the time, I thought there must be a better way to pull a train than with an on-board steam engine. No wonder that the railroads introduced electrification and diesel engines.

Lo and behold, some forty years later we were told that the future belongs to the steam engine. In 1968, a reputable automotive magazine was obliged to publish a roster of steam car enthusiasts from A for (Robert U. Ayres) to W (for Calvin and Charles Williams). It included professors from M.I.T. and other universities, and fancy pseudo-scientific names of the corporations founded by the enthusiasts.

The air pollution problem had, of course, triggered the revival of the Rankine cycle in the form of a steam or vapor engine. Burning fuel in a steam boiler offers all advantages for low exhaust pollution. It can almost be stated that fossil fuels would have to be abandoned altogether if they would form intolerable amounts of noxious combustion products, and that caused the enthusiasm for steam. But, and it is a big but, the clean burning is only one side, the bright side, of the steam engine. The other side is related to useful power derived from the burned fuel, and then things do not look so good.

Compared to a stationary steam powerplant, which permits incorporation of a number of fuel-saving components, a steam engine in a car or a truck has to be inefficient and could not provide tolerable mileage. The California steam projects with buses and cars proved that point conclusively.

William P. Lear (1902-1978), famous for his electronic inventions and his Lear Jet, was reluctant to give up. He told an SAE session: "Steam will become a success if one does it right." He thought a steam turbine instead of a reciprocator would be the answer. He subsequently presented his "Steam is Beautiful" bus at one of the coordinating meetings. I was privileged to ride in it and had a pleasant conversation with him and his charming wife. He already had conceded defeat, and was working on a supercritical wing and a composite airplane. I congratulated him; it was a better idea than dabbling with steam.

A few months afterward, an automotive company executive asked me to familiarize him with steam engines for vehicular applications. I replied: "It is very simple. It can be conveniently expressed by the number 2. Compared to a conventional engine, for a steam engine multiply the fuel consumption by 2, and divide the amount of exhaust emissions by 2."

To find an application for the vapor engine it was suggested that it could be utilized for converting the exhaust energy of truck diesel engines to mechanical power in a "bottoming cycle," thereby improving the fuel mileage. I countered this approach, citing that this would present an engine within an engine, with additional costs for its installation, service, maintenance, and reduced reliability. It would be far more prudent to explore the development and acquisition of better turbomachines to be used in turbocharged or turbocompounded diesel engines.

Though mileage improvements were demonstrated with bottoming cycles, the harsh realities of the marketplace prevented their adoption in land vehicles.

The Stirling Engine

The Stirling engine has a very checkered history. Invented in 1816 by the Scottish minister Robert Stirling, it was used in the middle of the last century, and, in limited quantities, as recently as in the 1920s.

A technical article on this "hot air engine" by the Dutch Philips concern drew my attention in the late 1930s. I have followed its progress ever since. Its thermodynamic process takes place in a closed cycle similar to the Rankine steam engine. As a working medium a gas is employed, either air, helium, or hydrogen. The combustion is continuous, and therefore offers advantages with regard to exhaust emissions.

Philips had revived the Stirling engine as a power source for remote radio stations, mainly because of its quiet operation. It built a number of small prototype engines and engine generator sets. The improvements in radio tubes and finally the advent of the transistor made the engine less attractive as a power source for small electronic installations. Philips therefore extended the aerothermodynamic, design and experimental studies to higher power levels and to various mechanical drive mechanisms. Fairly remarkable fuel consumptions were achieved. Experimental engines were installed in a yacht, a riding mower, generating set, a bus, and a compact car. Engine companies in Europe and in the U.S. entered into license agreements with Philips to explore and develop this powerplant.

Most conspicuous was the development by General Motors, who had accumulated 6500 hours of Stirling engine operation by the year 1965. However, GM dropped it in 1970 in favor of the Wankel engine.

The Philips company had tried to induce the U.S. Government to develop it as a passenger-car powerplant in view of its potential for both high fuel economy and low exhaust emissions. It demonstrated a Stirling engine-driven bus. After riding in it, I again had to refer to my Beier statement, "It moved!" Furthermore, one nagging item bothered me. Since its revival in the late 1930s, not a single application for the Stirling engine had been found nor an engine sold to a customer.

In 1972, the Ford Motor Company started to explore the Stirling principle as an automotive powerplant. JPL of Cal Tech hailed it as promising 30 to 45 percent fuel savings and practically pollution-free operation (see Appendix 11.2).

Ford discontinued its Stirling development in 1977. From then on, it was only pursued under government sponsorship. Recently, Stirling engines were installed in military pickup trucks. Multifuel capability is appreciated there, but the fuel mileage is only that of conventional engines, not better.

The Gas Turbine Engine

My 1946 monograph on the automotive gas turbine, the subsequent tank engine development in England, and my 1952 Mahle presentation "The Gas Turbine — Today and Tomorrow" kept me alert to worldwide gas turbine developments.

As an illustration, I cite two typical publications in prestigious engineering journals, one in England (the first one) and one in the U.S.

"And so we arrive at the present situation where, in spite of its triumphant advancement in one field of engineering, and in spite of its potential greatness in others, the gas turbine is as yet unknown to road transport. But the time is ripe for its introduction to this sphere and if we proceed through the thicket of initial difficulties we enter a new realm of engineering where the benefits are numerous and should go a long way toward satisfying the desires and hopes which engineers of road transport engines have held for the last fifty years. There is little time to lose if this country is to capture the markets of the world in what will undoubtedly be a new and huge sphere of industry. The immediate concentration of resources on what is undoubtedly a vital problem can give us the same technical superiority in this field that we have already established in the aircraft turbine field."

"Both Army and Navy engineers are currently saying that the gas turbine engines may well be on their way to competing with reciprocating engines as potential military vehicle powerplants. They are convinced that 'gas turbines can become generally competitive with reciprocating engines.'"

The British quotation of 1948 led to the first demonstration of a gas turbine-driven passenger car, the Rover "JET 1" in 1950; the American quote of 1962 led to the U.S. Army's 1965 RFQ for a tank engine and finally to the development of the AGT 1500 with production starting in 1979.

The Chrysler Corporation exhibited the first American gas turbine car at the 1962 New York automobile show. I was afraid they would prove me wrong with my dire predictions about the viability of a gas turbine-driven passenger car. But there was the glitzy turbine car! A P.R. woman was explaining the features of this new automotive powerplant, and praising its goodies. When she finished with the remark that for further details on the engine a development engineer would be available, I took this opportunity to ask him some questions, starting with: "Why does Chrysler think it can make a competitive gas turbine for an affordable price when all other companies had failed?" He replied: "We have various components which make that possible. For example, we do not use expensive superalloys, but ferritic steels, we use plated sheet metal in the combustor, carbon shaft seals, and so on." I said, "I know of those ferritics which exhibit high strength by adding ingredients like Niobium. However, if they reach a certain temperature, their strength falls off a cliff, and the engine is ruined." I also questioned him on some other claims. He finally realized that he could not properly answer my questions in technological terms, he therefore retreated with the question: "Do you really mean that our management would spend tens of millions of dollars on the development of this passenger-car gas turbine if it were not convinced of its bright future?" I ended the conversation with, "My dear fellow, I have seen worse things than that, and you should remember the fate of Ford's Edsel!"

I was wrong! A headline in the *SAE Journal*, June 1964, proclaimed, "Chrysler's Auto Turbine Travels The Long Road Home." Chrysler actually produced fifty gas turbine cars for evaluation on the road by selected customers. This mass demonstration was allegedly a great success. It did not lead to production, just to an award to its inventor for productionizing a gas turbine car.

In 1970, a prestigious California think-tank stated: "Of the many alternatives to the internal combustion engine (ICE), only the gas turbine shows clear promise of surpassing the ICE in terms of combined cost, performance, economy, durability, and low emissions. Regenerative turbines should capture over 10% of the heavy truck market by 1980. Turbines are likely to be offered by at least one foreign firm by 1973."

High specific power and low fuel consumption of the Brayton cycle can be achieved by raising the gas turbine inlet temperature. Metallic components in the hot section then require cooling which penalizes the performance somewhat. This becomes particularly detrimental with decreasing engine size. Ceramic materials eliminate this necessity. Despite decades of efforts, their brittleness prohibited application in gas turbine engines.

To enhance the technology of structural ceramics, a U.S. government agency initiated in 1971 a competition for a high-temperature automobile and an industrial gas turbine. I participated at Lycoming on its proposal, with cautious considerations in view of the programs' ambitious nature. A few ceramic turbine blades from my 30-year-old jet engine component collection exemplified my previous involvements in this field. Ford and Westinghouse won the contracts. Subsequently, one of our best engineers left for Ford to play a role in this fascinating work. I advised him on the uphill battle he will face, and the extremely slim chance that a Ford car will be driven by a ceramic turbine in the foreseeable future.

Parallel to this research project, General Motors and Ford tried their hands with metallic gas turbines for trucks, buses and industrial applications.

A Greyhound Super 7 Turbocruiser bus with a GM gas turbine entered revenue service in 1972 between Washington, New York and Boston. When it was on exhibit in nearby Bridgeport, I talked to its driver. He liked the vehicle when cruising on the highway, but was dissatisfied with its performance in the cities. Acceleration from a stoplight frustrated him the most. The passengers appreciated the smooth and quiet ride; they did not have to pay for the higher purchase, fuel and service costs.

In the same year, Ford started production of a regenerative gas turbine model in a factory in Toledo, Ohio. Advertisements called for mechanics to enter the gas turbine era "on the ground floor." In service, however, this powerplant was hardly up to par, thus forcing production stoppage after only 200 units had been built. The floor fell through, the turbines were recalled and the factory was quietly abandoned in 1973.

In the search for an alternative, pollution-free automobile powerplant, the gas turbine again emerged as a promising candidate. Chrysler continued its gas turbine passenger car development in 1971 with financial assistance from the government. The established targets for fuel mileage were fairly moderate. The 1973 oil embargo, with the ensuing dramatic rise in oil prices and concern about adequate fuel supply for the U.S., forced redirection of the program; higher fuel mileage had to be achieved.

This new obligation, ecology plus economy, presented an ambitious challenge to the automobile companies.

The Ford Motor Company wanted to take a wider view of the subject and asked the Jet Propulsion Laboratory (JPL) in November 1973 for a study of alternative engines. When that work with the title "Should We Have a New Engine?" was completed and released, I took the opportunity to comment on it (see Appendix 11.2).

Neither of the two JPL favorites, the Stirling engine nor the gas turbine, proved to be a practical alternative to the conventional engines. The gas turbine suffered from low fuel economy. This deficiency could be corrected only with higher turbine inlet temperatures. The panacea for this route was the application of ceramic materials instead of the expensive superalloys.

Again, a very ambitious government-sponsored program was established for this purpose. Contracts were awarded to Garrett/Ford, GM Allison, and the Williams Corporation. All three gas turbines operated at the high turbine inlet temperatures of 2350°-2500°F (1288°-1371°C), thereby promising excellent fuel mileage for their cars.

To stay in this international high-technology competition, the West German Federal Ministry for Research and Technology (BMFT) initiated in 1974 a program of ceramic components for automotive gas turbines to be conducted by the German motor vehicle and ceramics industry, research institutes and academia.

International conferences on the progress of this and other programs were held regularly. I remember a festive one in Washington, with a military flavor, attended by representatives of all NATO countries. A U.S. Marine Corps band entertained the guests; a color guard marched in displaying the flags of the participating countries. The speeches, slides and movies were exciting to hear and watch. If the hardware would have been half as good it would have been a real reason to celebrate.

Though ceramic gas turbine materials had been explored and investigated since the Second World War, they were not yet capable of fulfilling the stringent requirements of a passenger-car gas turbine. Reluctantly, the American programs had to be scaled back from car demonstrations to the development of ceramic components.

It appeared that I was wrong with the cautious prognosis to my friend! A 1977 magazine cover proudly depicted Ford's test run of a ceramic turbine with 2500°F (1371°C) turbine inlet temperature. With a sigh of relief, I found with a closer look at the data that the turbine power, speed and efficiency had missed the target by a mile.

At the social hour of another international "Ceramics for Automobile Gas Turbines" conference I was unable to cope with hors d'oeuvre, broccoli and a napkin in one hand and a glass of red wine in the other, with the result that the saucer fell to the floor and disintegrated into several pieces. Making the best of this faux pas I calmed the shocked audience with the words, "I'm not clumsy. I just wanted to demonstrate the real brittleness of ceramics and the problems confronting you!" By the way, I was not the first to focus on this deficient material property. Almost 500 years before, the Aztecs had apparently suffered failure of a ceramic turbine stator which even one of their gods in the hub was unable to prevent (see Figure 11.2). I had discovered this acute demonstration piece at the Museo Nacional de Antropologia in Mexico City during a Champion Conference.

Fig. 11.2. Aztec "ceramic turbine stator" (1490).

In the Fall of 1978, Ford decided to abandon the development of both an automobile gas turbine and a Stirling engine but to continue with the government-sponsored exploration of ceramic materials. Westinghouse had thrown in the towel years before.

To commemorate the automotive gas turbine, a $20 banquet was held in honor of its 25th anniversary in 1979. Accordingly, the birth date of the automotive gas turbine would be 1954, i.e., the public demonstration of our Parsons tank gas turbine. I doubt, however, that the sponsors had that event in mind. But what about 1950, the year of Rover's passenger car in England and Boeing's truck in the U.S.? It appears that the vehicular gas turbine eludes clear definitions both technologically and historically. The banqueters could purchase a Chrysler Turbine Car as a toy for their grandchildren or great-grandchildren who may be able to afford and drive a full-size gas turbine-powered automobile.

My sobering prognosis was also contested abroad, in Sweden and in Germany.

Sven Olof Kronogard (1916-1983) exhibited at a 1981 ATD meeting his KTT, the Kronogard Turbine Transmission he had invented in 1972. It employs a differential turbine principle. Sponsored by the Swedish government and the Volvo Co., it was developed for a Volvo passenger car. Planetary gearing and clutches interconnect the gas generator and the two power turbine stages. Improvements in torque characteristics and part-load fuel economy were claimed for this arrangement. Additionally, it led to a small powerplant package that was completed and tested in 1976. Due to Kronogard's interest in Lycoming's AGT 1500 tank gas turbine, I became acquainted and then corresponded with him. I admired the small size of the package, but refrained from expressing my opinion on its fuel economy. Starting in 1982, experimental cars were field tested with good results — and a better color movie. To repeat Chrysler's cross-continent run of 1964, one Volvo/KTT car was driven 2400 miles from Malmö via Copenhagen, Paris, and Nice to Monte Carlo. Among other honors, Kronogard was awarded a Royal Medal for his accomplishments in the automotive gas turbine field. With his untimely death in 1983, the project lost its enthusiastic sponsor and faded away.

Daimler-Benz of Stuttgart, the most active company in the field of alternative automobile engines, finally succeeded with the development of a ceramic turbine and installed it in a passenger car. Its name *Fernwagen* (long-distance car) indicated that it was considered more suitable for autobahn than for city driving. Though I enjoyed a standing invitation to drive this gas turbine automobile and passed through Stuttgart on several occasions, this unique opportunity eluded me for various reasons. Verbal descriptions of the turbine and the car were presented at the Dearborn CCM meetings but there are at present no plans for their production. On the contrary, DB just announced super sports cars with multivalve V-8 and V-12 engines.

Proponents of the ceramic gas turbine are still optimistic, and expect a small production run of some 800 to 1000 turbine-powered vehicles by the year 1995. In contrast, engineers who have monitored these government programs over the years claim that the lines of promise and the lines of achievement, originally far apart, are converging, but will not intersect before the end of the 21st century. In this connection, I once asked a friend discussing a similar situation: "Honestly, are you not overly optimistic?" To which he replied: "Of course, I'm getting paid for it!"

Ceramic parts found some use in piston engines for the reduction of friction, wear and heat losses. The latter property led to the "adiabatic" or better LHR (low heat rejection) engine. Progress is made in this field but the fantastic claims of an "adiabatic diesel engine" exhibited in an Army truck at an SAE Congress in Detroit were probably exaggerated, to put it mildly.

AGT 1500 in Abrams Tank

After my retirement, in loyalty to my former employer, I wrote an article for *WARD's Engine Update* (WEU, March 19, 1976)[5] headlined "Powerplants Also Duel In Tank Battle." I compared the two entries: the diesel, well-established and well-known by the Army and its generals, and the newcomer with a gas turbine that offered a few items in its favor, particularly half the bulk and half the weight of the diesel. (Its exhibit at the 1976 SAE Congress dramatically showed that.) I wrote further that this and the superior torque characteristic help to compensate for its slightly higher specific fuel consumption so that the cross-country mileage of the gas turbine tank may thus be competitive with that of the diesel tank. Other goodies are attributed to the gas turbine such as less field maintenance, longer time between overhauls, greater vehicle availability and, of course, less noise, no smoke and superior cold start capability.

Almost foreseeing Chrysler as the winner, I concluded somewhat over-enthusiastically with the statement that the boost for the gas turbine will spill over into the commercial field, first psychologically and then in real terms. The military paying for the costs of R&D, production initiation, tooling, and service evaluation will reduce the price for gas turbines so that users for specific applications may be able to afford it.

With the article's last paragraph I returned to the real world, writing that there is a sobering note of which gas turbine enthusiasts and optimists have to take heed. The success of the gas turbine in the air and now most likely on the ground is coupled to two ingredients: (1) high power concentration and high power, and (2) military aid.

---

[5] In view of the diminishing impact of the Wankel engine and the emerging alternative powerplants, Ward's Communications had replaced its WWR with WEU in July 1975.

Chrysler won the competition. On November 13, 1976, Lycoming was awarded the contract for its AGT 1500 with mass-production to start by the end of 1979.

Difficulties during the preproduction field testing were satisfactorily resolved. Lycoming shipped the first production unit in December 1979, 14 years after having received the development contract. It was a great event for all associated with the circuitous work on the AGT 1500!

Initial production and service problems cropped up, some so severe that a second supplier was suggested. The competition claimed with nasty newspaper ads to be well-equipped to fill the gap. The U.S. Congress declined for cost reasons to authorize a second source. Lycoming got its act together, and the M-1 Abrams Main Battle Tank became "the world's best tank."

One item, however, remained a nagging problem: the gas turbine's thirst for fuel. The M-1 tank consumed some 80 percent more fuel than its M-60 predecessor! This shortcoming could be only partly attributed to the higher power and superior performance of the tank. The deficiency prompted a fuel economy improvement program; it achieved an overall reduction of over 10% for a typical 48 hours of operation.

A continual disappointment also was the lack of another customer for the M-1 tank or AGT 1500 gas turbine. The Netherlands and Switzerland opted for the German diesel-driven Leopard II. In early 1990 the U.S. Congress finally dropped its opposition to sales to the Middle East: Saudi Arabia and Kuwait were to receive small numbers of M-1 tanks, and Egypt was to coproduce some. In August, the Gulf crisis precipitated changes in these plans.

To prepare for the next generation of a main battle tank, the Army instituted a study program entitled "Advanced Integrated Propulsion System (AIPS)." For fuel consumption and cost reasons, the diesel was to be included in the investigations. Of the proposals submitted by four companies for a gas turbine and by three for a diesel engine, the Army selected GE in 1984 for a recuperated two-shaft turbine and the Cummins Engine Co. for an oil-cooled, partly heat-insulated, highly turbocharged diesel. My previous concern about GE's competition was averted; Lycoming joined GE in 1987 in the development phase of the project. A competitive run-off between gas turbine and diesel is scheduled for 1990.

Proponents of commercial vehicular gas turbines had expected a much-needed boost from the M-1 installation. So far, it did not materialize. For me, this turn of events caused mixed feelings. On the one hand, I would have wished for my former Lycoming colleagues not only a temporary but a continuous success. On the other hand I had been less than enthusiastic about AGT 1500's intricate configuration; also, my Mahle paper was still on the mark.

The forthcoming testing of GE's gas turbine and Cummins diesel will be solely conducted with the powerplants in contrast to the tank competition five years ago when the powerplant represented only one item in the tank's assessment. The results, expected in 1993, will thus constitute a genuine all-inclusive comparison of the most modern specimens of a gas turbine and a diesel as propulsion systems for main battle tanks and other tracked vehicles. It will be exciting to watch whether the outcome will provide the definitive answer to a 50-year-old question of the best propulsion of a main battle tank. Will it be a diesel, a gas turbine engine, or none of the above, but as a wild card a DISC rotary?

Garrett/ITI Gas Turbine GT601

Developments of gas turbines for commercial vehicles were continually conducted worldwide but none had led to production. The Garrett Corp. was successfully developing and marketing gas turbines in the low-power class and dominated the APU market (auxiliary power units). It also had been active in heat exchanger technology.

With this promising background, it wanted to enter the commercial vehicular gas turbine field. In 1972, a consortium of four major engine and heavy truck manufacturers was formed. This Industrial Turbines International (ITI) was jointly owned by two U.S. and two European companies, the Garrett Corp. of Los Angeles, CA, Mack Trucks, Inc., of Allentown, PA, AB Volvo of Göteborg, Sweden, and Klöckner-Humboldt-Deutz (KHD) of Cologne, West Germany. (Volvo parted with the project in 1975.)

The ITI gas turbine engine, Model GT601, with a power of 450-650 horsepower, consisted of a single-spool gas generator with two radial compressor stages, driven by a radial-inflow turbine, a novel recuperator and a single-can combustor. The free power turbine with two axial stages and variable stators delivered power via a reduction gear to the vehicle transmission. Power transfer to the gas generator also was provided. I was happy to notice the similarities to our Parsons tank gas turbine of a quarter of a century ago.

In 1974 satisfactory progress with the engine tests prompted a request for a second source of the engine control system, particularly for the European market. In this phase I acted as a consultant and thus became familiar with the project.

The powerplant was field tested in the U.S. and in Germany, fulfilling the initial fuel consumption and exhaust emission targets. Some redesign made the manufacturing costs competitive. Glowing reports of this new propulsion system for heavy trucks appeared in the trade press. Road mileage was allegedly equal or better than that of diesel-engined trucks. My own rudimentary analysis indicated that claim to be debatable, particularly in view of the idle fuel flow which was three to four times that of an equivalent diesel. Also, turbocompounded diesels had demonstrated a 15% mileage improvement.

I again saw the engine at the 1979 International Auto Show in Frankfurt, proudly exhibited as a neat package. However, its power was apparently too high for the European market. For North America, it also was required only for acceleration and hill climbing. In short, commercial customers were reluctant to install the engine.[6]

With the successful application of Lycoming's AGT 1500 in the Abrams tank, the military route appeared promising. A variety of field tests with tracked vehicles demonstrated the engine's superior capabilities. The reliability, durability and long service life also were appreciated but did not lead to production. After fifteen years of gallant efforts in the commercial and military markets, the project faded quietly.

My inquisitive mind again raises the old, nagging question: If the ITI 500-horsepower gas turbine was unable to hack it for trucks, and Lycoming's 1500-hp becomes noted for its low fuel economy in a tank, why would people dare to promise that they can produce within a few years a 100-hp gas turbine that can compete in the passenger car market?

Electric and Hybrid Vehicles

In the early 1960s, interest in electric vehicles was revived for a variety of reasons such as beneficial spin-offs from space efforts and the opportunity of electric utilities for load leveling by selling electricity at night.

Storage batteries had improved, but hardly enough to become competitive. Fuel cells appeared to offer a viable alternative. I vividly remember having seen an electric motorcycle with an on-board fuel cell in New York City. Also, at Curtiss-Wright we were swamped with fuel cell suggestions. The promoters with their often wild predictions considered us "square" if we ignored them. Over 100 companies worldwide actively pursued fuel cell development. This technology succeeded in space vehicles, found some niches as small power stations, but failed in automotive applications.

The 1970 Clean Air Act provided another impetus for the electrics: the federal government-sponsored R&D programs. The 1973 oil embargo prompted some opportunistic entrepreneurs to mass-produce electric cars. That spurt was of short duration, though.

---

[6] Some ten years later, Caterpillar and Cummins started to offer truck diesels with the highest rated power of 460 hp.

229

To keep abreast of the government programs, I attended some contractor coordination meetings. Once the proceedings were interrupted with the announcement that an advanced lead-acid battery had exceeded an energy density of 40 watt-hours per kilogram. I failed to grasp the significance; that figure was still inferior to the fossil-fueled internal combustion engine by a huge factor. In his book Fundamental Terms for Automobiles, L. Baudry de Saunier drew attention to the battery problem with these words: "The efforts toward improvements of electric vehicles have to be directed to the batteries, but only a thorough revolution may yank them from their old tracks." Noteworthy is the year of the book's publication — 1902!

The development of electrics culminated in the second half of the 1970s and first half of the 1980s. Organizations, councils, publications, exhibitions (Drive Electric-'80), symposia, etc., sprung up all over the world. I once stated that if the progress of electrics would keep only a fraction of the pace of their publicities, they would have made it.

One decision illustrated the situation more than any other. Mail delivery by the U.S. Post Office would certainly constitute an ideal application for an electric vehicle. Experimental vehicles were field tested; however, when the Post Office issued an RFQ for 100,000 delivery vans, electric and diesel-driven vehicles were excluded.

In 1979, the government programs resulted in the display and drive of a prototype electric, seating two adults and two children. This milestone was introduced with the prediction, "This demonstration will finally lay to rest the outmoded opinion of electric passenger cars!" As an exhibit, it was an impressive sight. After a test drive, however, a reporter characterized the car differently: "Slow as molasses," he said. An almost identical car was demonstrated and promoted in Germany in 1986. In a letter, I compared it to the above demonstration and asked which means had eliminated the "molasses syndrome." I was told privately that the car had no difficulty in adjusting to the prevailing city traffic pattern. (And that in Germany!) As of this writing, I am unaware of its commercial success.

To overcome the lack of driving range and power, hybrids were suggested and demonstrated. They comprise an internal combustion engine and a battery/electric motor system, i.e., the best of both worlds. It worked fine, but not as expected. During driving, one of the power systems constitutes a deadweight and negates the advantages it offers when in use.

The battery development needs a breakthrough similar to that which the transistor and/or chip had provided over the vacuum tube. Such a giant step had been predicted every other decade, but it failed to materialize. It appears that power-producing electrons require lots of space (and thus weight) and/or higher than ambient operating temperatures.

Electric cars were promised by Ford in 1960, by GM in 1977 and again in 1990, incidentally in each case by the retiring president or chairman. Outsiders entered the field, too. In 1980 Gulf + Western heralded an "Electric Engine" with zinc and chlorine as the working mediums, demonstrated experimental cars, and drove them over tens of thousands of miles. Something must have gone wrong, though; the project was scuttled in 1983. The English single-seat three-wheeler Sinclair C5, a derivative of electric leisure vehicles, eased the battery problem by providing an additional pedal drive for acceleration and hill climbing. It found insufficient numbers of buyers and production had to be stopped. A Danish "Whisper" electromobile suffered a similar fate.

The high-temperature approach received renewed attention, particularly in Europe.

A jolt similar to Gulf + Western's of the early 1980s recently arrived from Japan. Isuzu claims a revolutionary battery with not only fantastic properties but the capability of being recharged in 30 seconds. It reminds me of my days at CW when I was inundated with fuel cell claims.

The electric vehicle field may be assessed as follows: It will not be a question of what an electric car will offer in performance, acceleration, range, and overall costs, but how little of these customary characteristics the motoring public will be willing to accept for the sake of a cleaner atmosphere. It may take laws to wean the public away from the customary, economic and reliable ICE-driven automobile; smog-plagued states will be first to pass these laws.

Energy Storage

CW had pursued projects which in my view were rather exotic, offering only slim chances for success. Then the rotary engine era had spawned a variety of engine concepts, some of them farfetched. Their descriptions and assessments are included in my Detroit Library collection as curiosities.

The pursuit of very strange concepts in the ongoing search for economic and ecological automotive power systems caused me to think "here we go again."

One fascinating idea involved the mechanical storage of energy. Viable solutions to the age-old energy storage problem exist for power stations in the form of storing water at higher elevations, air at higher pressures, electricity in batteries, and steam or hydrogen in pressurized storage vessels. Mechanical storage had been tried in an electric city bus incorporating a big flywheel. An electric motor, supplied with current at a bus stop, brought it up to high speed which provided the drive energy to the next stop. The economics of this system suffered due to the low ratio of stored energy to weight of metallic flywheels. Space-age technology was supposed to offer extremely high ratios with super flywheels made from high-strength fibers. The race was on with companies specializing in this field.

Flywheels were designed for load-leveling of power stations, for storing and reusing the braking energy of ordinary vehicles and of hybrid electric cars, for a commuter car, etc. The latter application was a particularly bold and ambitious project. A small car contains a super flywheel providing the drive energy sufficient to commute to work. At home, an electric motor recharges the flywheel at the cheap night rate.

A promising application for flywheels seemed to be subway trains in view of their frequent braking, stopping, and acceleration. In regular service in New York City, an experimental system had reduced the costs for electricity and for brake shoes by high amounts. In view of operational and safety problems, however, the system was not generally introduced.

The other applications failed to succeed beyond experimental stages. I believe that the commuter car already succumbed on the drawing board.

Appendix 11.1

The Wankel Fever as Expressed on the Cover and in Headlines of Magazines

Quick (1960)
> The Thirty-Years War of Felix Wankel: 1930 Outsider, 1959 Victor. The automobile invented for the second time. On the threshold of the production line.

Der Spiegel (26. Juli 1961)
> Engine inventor Wankel. NSU partner Hurley: Icebreaker, crook, or genius?

Popular Science (April 1966)
> "Wankel fever" hits Detroit. First test drive of U.S. car with Curtiss-Wright engine RC2-60. President T. Roland Berner shares with his engineers a pride of accomplishment.

Machine Design (January 7, 1971)
> An idea whose time has come? The Wankel revolution.

The New York Times Magazine (... 1971)
> The little engine that could be an answer to pollution.

Popular Science (100th Anniversary Issue, May 1972)
> You're witnessing the start of a revolution in the auto industry — GM Rotary Engine for the '74 Vega.

Ward's Wankel Report. New biweekly magazine started July 12, 1972.

Fortune (July 1972)
> A car that may shape the industry's future.

Changing Times: The Kiplinger Magazine (July 1972)
> Is the Wankel the auto engine of the future?

Iron Age (August 8, 1972)
  When Wankel puts it all together — watch out.

Mazda advertisement (TIME, October 23, 1972)
  The "elegant" engine.

Motor Trend (November 1972)
  Special: The rotary engine. The engine Detroit couldn't ignore. Interview with Dr. David Cole. Rotary round-up; from recreational toys to weapons of war, it's a rotary world.

Popular Imported Cars (January 1973)
  The whole Wankel story. Mazda receives our certificate of excellence award.

Road & Track (November 1973)
  Mazda GT — One man's dream car.

Motor Trend (December 1973)
  The hot 4-rotor Corvette is here.

Motor (September 28, 1974)
  Whither the Wankel? Who's staked what in the great Wankel gamble? Thirsty or thrifty. Wankels on the market... and others not for sale.

Fast Lane (January 1976)
  Round in circles. The three "nursemaids": Dr. Walter Froede of NSU, Dr. Max Bentele of Curtiss-Wright, Dr. Kenichi Yamamoto of Toyo Kogyo (Mazda).

Appendix 11.2

Comments on JPL's Study "Should We Have a New Engine?"

Excerpts published in Ward's Engine Update, 11/11/75 under the headline: Bentele: JPL Report Is "Superfluous"

The Ford Motor Company had pursued in search of a better idea the orthodox Otto-cycle automobile engine and its stratified-charge derivative, the Brayton-cycle gas turbine in one weird and some more conventional forms, the Rankine-cycle steam and the Stirling hot-air engine. After having unsuccessfully tried to put a truck gas turbine into production and having cancelled the Rankine, Ford retrenched to a high-temperature ceramic turbine (with Government financial assistance) and to a Stirling (with the help of the Dutch Philips concern). Time for reassessment was due whether that was the right track now. Probably for fear of antitrust violations, the Ford Motor Company went as far away from the automobile industry as geographically and technologically possible, namely to California and into space. In November 1973, JPL of Cal Tech was awarded a $500,000 contract to study the automobile engine field.

On September 4 of this year (1975), I read brief accounts of the Study. Lo and behold, it said: There is still a bit to be squeezed out of the 100-year-old Otto engine; the stratified charge and Diesel are dead-end streets. Forget about the Rankine. Combined cycles are too difficult to handle. Electrics and hybrids are a nuisance and should go away. The Wankel? Don't mention it.

However, the Brayton and Stirling promise 30 to 45% fuel savings, are practically pollution-free and can be mass-produced by 1985 or earlier at a cost differential small enough to be easily recovered by the first owner in fuel savings. Only an aggressive development program and firm commitment was required to achieve this goal. It was so simple.

Unimpressed, I underlined a few passages, scribbled on the margin "Des' Brot ich ess, des' Lied ich sing" (which means "I'll sing the song of the one who feeds me" or in short "I'll toe the party-line") and put the clippings in my folder in which I had collected numerous studies and similar predictions over the last 30 years. That was a mistake!

The Study caused quite a stir due to JPL's prestige in aerospace, its sheer size (1-1/2 inches thick), time to completion (18 months), or any other reason. The SAE organized a session for the investigators to present their case and for a panel to discuss it. My spirit was lifted when it was said that only real but not paper engines will be introduced, but it sagged again when I heard about scenarios and strategies reminding me of those in the Pentagon Papers which did not turn out too well.[1] From my days in industry I remember the definition of a paper engine as follows: an engine mechanically designed in fair detail on sheets of paper, its performance based and substantiated by applicable test data of the elements and allowances made for inherent losses occurring between those elements and for parasitic losses caused by maldistribution of the gasflow, thermal distortions, leakages, non-optimum flow passages for mechanical, vibrational, manufacturing, packaging or any other reasons, etc. The performance predictions are usually made better to suit the purpose of the paper engine, that means a lot better "to get a foot in the door" or just a little or no better if a contract is expected which would include a penalty clause for missing the performance.

Unfortunately, neither real nor paper engines were presented but mere engine configurations. Their basic components were described in general terms and attached with optimistic performance numbers gleaned from sources within the Free World. These in combination with dubious allowances for losses naturally led to the promised fuel savings.

The Technical Reports were equally disappointing. The described configurations may be converted into aerothermodynamic gadgets or robust engines pending on the ingenuity of the design team and its leader. Most signs in the Report were in the direction toward the former endproduct, whereas, of course, only the latter would survive in the marketplace. The Report shows, admittedly in the greatest detail, what has to be accomplished but not how. And the what is known since 1944 in the case of the Brayton, for example. The text is more educational than constructive. Also, some wizardry was employed to make the favorites really attractive.

Apart from the debatable baseline car, let us consider one example. A single-shaft Brayton requires some 30% less power than its Otto-engine equivalent car. (The vehicle top speed is thereby reduced but still adequate, it says.) Sure, that

---

[1] Robert S. McNamara, then Secretary of Defense, had commissioned in 1967 a top-secret history of the United States role in Indochina. In 1971, the *New York Times* had obtained most of the narrative and documents. After some legal hustles, it then published them as "The Pentagon Papers."

substantial power reduction is possible by the higher power/weight ratio of the propulsion system and, consequently, lighter car. Wrong! The Stirling car weight is roughly equal to that of its equivalent but its power is some 20% less. It is the torque-speed characteristic of the propulsion system, it says. Now then, this characteristic of a free-turbine Brayton is most suitable for a car (Rover originally proposed two gears only, one for steep hills and one for ordinary driving), the single-shaft Brayton is the worst and the Otto is in-between. So how come the Otto requires 40% more power than a single-shaft turbine?

Are there more questions and missing links? Quite a few: not nearly enough considerations on size effects and its implications, on off-design and transient conditions, on operational emergencies and safety (what happens when the control system goes haywire? You are probably safe but may have to buy another engine), on service, maintenance and repair.

So the JPL men failed to present a convincing argument for their selections and recommendations? Yes, by their own admission. They feel to have greater luck with politicians than with businessmen. Their calculated development cost for their three favorites would amount to .8 to 1.2 billion dollars over a period of 5 to 8 years. They therefore expect cold shoulders from the automakers despite their promises of a more expensive car (meaning higher profits), a pollution-free car (Congress will be off their backs indefinitely) and tremendous fuel savings (the public will stand in line for these cars). Let me remind them that General Motors has and will spend 2 billion dollars from 1967 to 1977 on pollution research alone (letter President E.M. Estes in Wall Street Journal 2/24/75). Why should GM not grab their proposal? It can only be surmised that they are probably aware of occasional differences between predictions and the real world. And unwittingly, JPL supports this argument. Their contract called for a report within 14 months, the Report quotes an 18-month study, and it took 21 months to deliver it.

Now some specific exceptions to the conclusions and recommendations:

1. In a free society it is not a good idea for the Government to dictate to the engine designer or to restrain his freedom. Mild examples are the British tax formula based on piston area which led to heavy long-stroke engines and the German formula based on displacement which led to sensitive high-speed engines.

2. A Government-supported crash program on two engine types would bring all other efforts toward viable solutions to the fuel economy/ecology problem to a standstill. It would also make the U.S. vulnerable to possible imports.

3. The Otto recip and Wankel engine have still a lot left in them. As an illustration, EPA mileage figures show the best cars in their respective size classes some 20% better than the average and some 50% better than the worst ones.

Everybody being so eager to give advice to those who have to do the work and to those who have to pay them with their own not somebody else's money, we wish to join the parade, too, and offer some suggestions:

1. Establish an honest dialogue between the scientific and engineering community, auto industry and Government aiming at well-substantiated long-term emission standards.

2. Plug away toward the best fuel economy/emission car on all fronts. When JPL's favorites accomplish a milestone the popping of champagne bottles will be heard around the world; it will be used to reassess other programs.

3. Let the Government assist industry by component research and/or release of data from military programs that would not jeopardize our national defense such as from the Army's STAGG program (Small Turbine Advanced Gas Generator).

4. Finally, let the free market decide and let the fittest harvest the fruits of his labor.

In summation then, is the JPL study right or wrong? Neither, it was superfluous.

Chapter 12

Engineering History, Collections, Rights

In October 1971 the University of Wyoming invited me to consider placing my papers (correspondence, logs, diaries, journals, manuscripts, photographs, books and other literary memorabilia) in their Transportation History Foundation, a collection "pertaining to the history and development of aviation and aeronautics in the twentieth century." I felt highly honored by this invitation, but had some reservations because a major portion of my professional career was dedicated to engines and vehicles suitable for ground transportation, and specifically to the Wankel engine which deserved special treatment. I therefore asked a Detroit engineer, whom I had met during the Wankel engine development at Curtiss-Wright in the mid-1960s, for advice. He suggested the Automotive History Collection of the Detroit Public Library. This collection encompasses areas of road transportation only, but constitutes the most complete set of records on the subject in this country. After a visit with James J. Bradley, the head of the Automotive History Collection, to see the collection and discuss the invitation, I was happy to place that portion of my papers and documents related to the automotive field in this Collection, with the understanding that the University of Wyoming would receive the papers dealing with aviation.

During the preparation of the collection, some events influenced my personal judgment on the type of papers to be placed there initially and contributed to my delayed contact with Curtiss-Wright on the planned establishment of my collection.

In November 1971 the Chilton Book Company of Philadelphia announced the book, The Wankel Engine. Design. Development. Applications, by Jan P. Norbye, with the claim that the author "had spent years collecting information on the Wankel, and discussing it with engineers involved in its development." The latter statement bothered me. Though Norbye is, as I am, a member of the SAE, he never contacted me or talked to me. The first printing of the book did devote substantial attention to the role played by Curtiss-Wright and by me. However, by a combination of omissions and misstatements, it did not accurately or adequately present my contributions. Pictures and paragraphs of my SAE papers and others were included in the book, but the book's bibliography did not include any of these, nor my name. Up to the publication of the book, I had published seven papers on the Wankel engine in American and foreign journals. At that time my SAE papers carried the statement, "For permission to publish this paper in full or in part, contact the SAE publications division and the authors." Neither the Society nor myself had been contacted or had given permission.

Fortunately I have two lawyers in my family, my daughter Ursula and her husband, Buzz. With their assistance, I communicated my objections to the Chilton Book Company, which finally agreed to put our desired corrections into the second printing of the book, and did that to my satisfaction in August 1972.

An additional benefit of this affair was SAE's review of its copyright situation. I was happy that SAE introduced a legal copyright note on its papers if the author wishes that.

This Norbye book and other events prompted my resolve to establish my fundamental contributions to the rotary engine and to protect my professional reputation. The invitations for collections to the Transportation History Foundation and the Automotive History Collection presented unique opportunities to accomplish this goal.

In May 1972 I presented the "Max Bentele Collection of Wankel, Rotary and Other Engines" to the Automotive History Collection of the Detroit Public Library (see Appendix 12.1). The contents of the folders are listed in Appendix 12.2. The Collection includes descriptions and my appraisal of a variety of novel engines and is supplemented with engine proposals which I have reviewed and assessed since then.

The presentation of my collection was favorably reported in the general and professional press, both in the U.S. and abroad. I also received appreciations and congratulatory letters from friends who had worked with me on the development of the Wankel engine. A German company, a leading developer and manufacturer of gas sealing components for piston and rotary engines, wanted to buy the collection. Obviously, it is not for sale.

Over a year later, Curtiss-Wright tried to retrieve those parts of the collection which were, in their opinion, Curtiss-Wright property. This request surprised me. I had meticulously avoided including those papers or documents that might be considered trade secrets or would jeopardize filing of patents and/or negotiations with prospective licensees of the engine. Also, some of the papers in question were 14 years old, i.e., confidentiality had, in my opinion, expired.

I asked the professional societies to which I belonged (SAE, AIAA, Sigma Xi) for an opinion on the subject and their possible assistance. They all politely declined to take a stand. It was a question between the employer and the employee, I was told.

In November 1973 I resolved my dispute with Curtiss-Wright amicably and my Collection was again opened to the public, in full.

Some ten years later, the U.S. Congress debated a similarly important matter, a bill to define the employer's obligation toward employee's inventions. The SAE considered such a law uncalled for. I expressed my opinion in "Feedback" of *Automotive Engineering*, May 1983. I commented on some of the issues raised (the issues are cited between quotation marks). "The inventor is employed to invent and is fairly compensated for this work." The latter may be true for many corporations but certainly not for all. In addition, if corporations with high ethics and character in this area change management, merge, or are sold, the new situation may be unfavorable to the inventor. Without legal protection of his rights to his inventions, only one right is left to him, namely to quit. Some corporations proudly present dollar figures of obtained royalties in their quarterly and annual reports, but think of their own inventors last, if at all.

"The proposed law may ignite an adversary relationship between employer and employee." This adversary relationship is inherent in the status quo where the employer possesses all legal rights to an invention and the employee none. Clarity promotes good relationships; ambiguity subverts them!

My Collection to the National Automotive History Collection[1] raised another question. Such contributions to public institutions are tax deductible. Politicians, authors, artists and other public figures took advantage of the existing law, and claims for their collections with five- or six-figure dollar amounts had been allowed by the Internal Revenue Service. This "pork barrel" loophole was subsequently closed; only the "market value" of contributions is now allowed. It is a fascinating question as to which appraiser determines that figure and who appraises the qualification of the appraiser.

In my above-mentioned "Feedback" letter, I state, "For a long time, a major concern of mine has been to improve the status of the engineer and the inventor in our society." I still feel the professional societies have an obligation to address the employee/employer relationship with regard to the questions I had experienced and had to deal with on my own, without their assistance. The same duty also applies to collections; otherwise valuable documents will be lost forever.

---

[1] In 1975 the Detroit Library Collection became "National" and Bradley became its Curator.

Appendix 12.1

Presentation of the "Max Bentele Collection" of Wankel, Rotary & Other Engines to the Detroit Public Library's Automotive History Collection

Monday, May 22, 1972

James Bradley, Chief of the Automotive History Collection, welcomed the audience on behalf of the Library, expressed gratitude for receiving the Collection, and then introduced Dr. Max Bentele who donated his Collection with the following remarks:

Ladies and Gentlemen:

There are two reasons for this Collection, an objective and a subjective or personal one.

The reciprocating internal combustion engine dominated the automotive field for almost a century. It is now being challenged by rotary and gas turbine engines as well as external combustion engines such as steam and Rankine-cycle engines.

Rotary engines have been invented concurrently with the reciprocating engine; James Watt and his associates already had one running. However, as internal combustion engines they are new on the scene and only one has succeeded to the development and production stage. That, of course, is the Wankel Engine, invented in 1954 by the German self-made engineer Felix Wankel. The first experimental Wankel engine in its present known version ran in 1958, and in the meantime more than 300,000 have been produced and sold. It is thus quite natural that this

245

engine is being discussed in technical and popular journals and magazines, and that books are written about it. This Collection is intended to supply source material, some published and some unpublished, to the development history of this engine and to bring it into perspective to other rotary engines which at one time or another had made headlines but did not make the grade. Examples are the old Cooley Engine, promoted by Renault in France and American Motors in the U.S., the Mallory Engine, a vane-type engine tried and then dropped by a famous aircraft engine manufacturer, and the Marshall, still being publicized by its inventor but to the best of my knowledge not even having run anywhere. These are just a few, more you will find in the Collection with my evaluation and remarks.

Now to the personal reason. Most of my professional career was devoted to the internal combustion engine in the form of four-cycle and two-cycle engines, jet engines, and gas turbines. In early May 1958, the then president of Curtiss-Wright called me to his office, handed me a crude plastic geometric model with the words: "That is supposed to be an engine. Find out whether it is, how it works, what are the pros and cons, advantages and disadvantages, problems, etc. Do it at home over the weekend and don't talk to anybody about it." Well, I got to work, found that it is an engine operating on a genuine four-stroke cycle with the advantage of a simple two-cycle, no valves and valve mechanisms, that it will be small and light, and without vibrations. These were the positive sides. Now the problems: number one was the gas sealing. The combustion chamber at peak pressure had to be sealed against the two other chambers and the crankcase the same as in a reciprocating engine. However, whereas there the seals are in a round cylinder and put in series so that the full pressure doesn't act on one alone, this engine had to have a complex sealing path and the apex seal could only seal with one line conduct. This problem looked tough but not insurmountable. The other problems were found to be of a lesser degree.

I reported back on these findings and was told then about Wankel as the inventor and NSU as the developing company. I gave my whole-hearted nod to the engine not because I knew most of the German people involved but on the basis of my purely technical and engineering analysis. This position of mine was so decisive that Curtiss-Wright obtained the rights to the engine which, in turn, assured its development for which, on this side of the Atlantic, I was put in charge. I fulfilled this task, with the help of my colleagues, with full success. When I left Curtiss-Wright in 1967, their Wankel engine was ready for production.

The Collection records the whole early technical history of the engine, the first public announcement in November 1959, the Munich Symposium of January 1960, the first SAE Paper of March 1960, and my papers and lectures on the engine, the highlights being the Rotary Engine session of the SAE International Congress and Exhibition here in Detroit in January 1961, and my paper delivered in Milwaukee where the Outboard Marine Corporation has now started field service testing of their Wankel engine.

So it took a long time, much hard work, many headaches, but also joy, from the plastic model to the recent announcement in the *Wall Street Journal* that General Motors allocated 50 million dollars over a period of five years for obtaining the rights to this engine.

This Collection will give a lot of background to this success story. It will be of value not only to the historian, but also to the student and engineer who will find many examples of sound engineering thinking, which is relatively cheap or not well rewarded in contrast to spending lots of development dollars in the hope something will come out of it.

I am happy and it gives me great pleasure to present the "Max Bentele Collection" on Wankel, Rotary and other engines to the Automotive History Collection of the Detroit Public Library. Thank you for your attention.

The files of the Collection were laid out on tables as well as the List of Contents. Interesting highlights were displayed for perusal by the attendees of the ceremony. Writers and engineers also took to opportunity to talk about specific items with the donor.

The next day the Chairman of the Rotary Engine Session of the SAE Congress and Exhibition in the Detroit Cobo Hall introduced Dr. Bentele to the audience of over 500 engineers and scientists, and drew their attention to the Collection. These announcements were greeted with great applause.

The donation was subsequently reported in the general and professional press both in the U.S. and abroad.

Appendix 12.2

The Max Bentele Collection

1) Rotary and Other Engines

    a) Surveys (F. Huf, F. Sisto, etc.)

        "History of the Rotary Piston Machines," by F. Huf, treatise published by *Automobil Revue*, 1961, in German and French, comprehensively covering the period of 1588 to 1959.

        "Comparison of Some Rotary Piston Engines," by Prof. F. Sisto of Stevens Institute of Technology, contractual investigations summarized in SAE Paper (1963), with discussion.

        "Rotary Combustion Engines," article in three parts in *Automobile Engineer*, by R.F. Ansdale, 1963-64.

        Geometry of Hypotrochoidal and Epitrochoidal Machines; Maillard, Sensaud de Lavaud, Cooley, Wankel.

        Miscellaneous rotary engines described in technical magazines.

    b) Sensaud de Lavaud

        French patent #853,807 (1939-40) evaluated as forerunner of Wankel engine. Correspondence with president of French company verifying the evaluation.

    c) Cooley-Renault-American Motors

        Cooley patents (1903) of Cooley Engine Company in New Jersey which manufactured steam engines of this type. Comparison with Wankel engine; Wankel far superior.

Announcements (1963) by Renault and American Motors on their development work on this engine type to "beat the Wankel." Thirty sheets of Renault patents indicating the substantial effort spent on this engine. Development discontinued.

d) Mallory

Vane-type engine published in *Aviation Week* (1967) as being superior to the Wankel engine. M. Bentele's evaluation of engine patent and description negative.

Pratt & Whitney Aircraft took option on a license of the Mallory engine, could not make the experimental engine operate and finally dropped the project.

e) Lundquist's Dyna-Star

A piston engine with rotary engine features, covered by patent #2,989,022. Evaluation and comparisons to other engines summarized in memorandum.

Engine named "Dyna-Star" is fully described by patent, SAE Papers, and articles in technical magazines (originals).

Thiokol Chemical Corporation was in process of developing the engine from 1963 to 1968 as documented by articles and four brochures (originals).

Engine probably not pursued further.

f) Marshall's Tri-Dyne

True rotary engine incorporating three rotors without rubbing seals. Technical descriptions and high claims in *The Engineer* (1968) and *Popular Science* (1969). AVCO Lycoming investigated the engine (called Roto-Lobe) but concluded by M. Bentele's final evaluation — feasible but not competitive.

g) Kauertz

In 1962, well-publicized engine (McGraw Hill, *Popular Mechanics, Hobby*) as competition to Wankel. Evaluation negative. Engine not in the news anymore.

h) Mercer

Publicized in 1969. Evaluation negative. Engine probably not pursued further.

i) Miscellaneous Engines

Descriptions and patents of a variety of other rotary engines (8), exemplifying but by no means comprehensive. Some high claims of engine superiority but few engine concepts reaching the development and none the production stage.

j) Further Engine Evaluations

Some engines did not go beyond the drawing board or patent office; some made it to experiments on the test stand; some were demonstrated in vehicle and other applications; none, however, achieved production.

The list in alphabetical order of the inventor:

Arajo; Bates; Dean; Ehrlich; Erickson; Giesel (gas turbine/diesel); Gilbert; Heydrich; Hinckley/Beloit/Hornbastel; Hovorka; Huf; Katobi (Stirling derivative); Marin; Murray; Null; Ruf; Ryen; Sarich Orbital; Smith Subira; Waller; Walter.

As time goes on, the list will grow and grow.

2) Wankel Engine

a) First U.S. evaluation by M. Bentele

b) Early designs

c) First announcement in English

d) Early brochures

e) Geometry/displacement

3) Wankel Engine — Technical Papers

4) Wankel Engine — General Publications

5)   Wankel Engine — Patents

6)   First Wankel Engine Symposium — Munich 1960

7)   Wankel Engine Sealing

8)   NSU Spider — First Production Car (with Wankel Engine)

9)   NSU Ro 80 (with Wankel Engine)

10) Toyo Kogyo — Wankel Engine Developments

11) Fichtel & Sachs — Wankel Engine

12) Mercedes-Benz — C111 Wankel Engine Car

13) General Motors — Wankel Engine Developments

14) M. Bentele's Papers and Lectures

   a)   SAE International Congress & Exposition, Detroit, Jan. 1961

   b)   SAE, The City College, New York, April 26, 1961

   c)   SAE, Milwaukee, October 6, 1961

   d)   ASME, Penn State, February 1962

   e)   SAE, Twin City, March 21, 1962

   f)   Yale, Nov. 27, 1962/MIT, Feb. 25, 1962

   g)   Princeton University, Edwin G. Baetjer II Colloquia, Oct. 31, 1963

   h)   ASME, Stevens Institute of Technology, April 1, 1964

   i)   Stevens Institute of Technology, Applied Mechanics Seminar, Dec. 17, 1964

   j)   SAE, Baltimore, February 16, 1967

   k)   SAE, Miscellaneous

Into life's ocean the youth with a thousand masts
daringly launches;
Mute, in a boat sav'd from wreck, enters the
greybeard the port!
Friedrich von Schiller (1759-1805)

Biography

I was born on January 15, 1909, in Jungingen, a small village of farmers, crafts and tradesmen, and industry workers. (In 1975, Jungingen became a borough of the University City Ulm.) My father Carl Christian and my mother Walburga, née Ruhland, were both born in 1869 and married in 1895. They owned and operated a small farm and a workshop for harnessmaking, upholstery and home decorating. My father, a certified master of his trade, actively participated in community affairs. He was elected to the Village Council, and served as treasurer of the local savings bank and of the church. Furthermore, he acted as official meat inspector and as a veterinarian assistant. I was the youngest of six children, with three sisters and two brothers.

One of my earliest recollections is the start of the First World War, with the conscription of men and horses. Alarming rumors, fear of spies and of an invasion prompted the villagers to set up makeshift road barriers.

The war brought many hardships to us, but hunger was not one of them. My parents, brothers and sisters were keen to teach me reading and writing to provide me with a head start in school. At the age of six, I entered the local elementary school. In 1917, I transferred to the *Realschule* (secondary school) in Ulm, four miles from home. I felt great wearing the traditional student cap (Figure B.1). My transportation was by train, bicycle or on foot, depending on the weather.

My brother Carl, drafted in 1915, was wounded in the Battle of the Somme. After his recuperation in an Ulm hospital, he was sent to the eastern front. My brother Ernst was delegated to the Balkan theater in 1917.

The November 1918 revolution ended the war on the military battle fronts, but not at home. I saw machine guns in the streets and squares of Ulm, shattered shop windows, and demonstrations by the workers. For the first time I heard the battle cry of the socialist Internationale, "All wheels of industry will be standing still if your strong arm exerts your will!"

Fig. B.1. In my school attire at the age of eight (1917).

Fortunately, my brothers returned home before Christmas, Carl from Taganrog on the Sea of Azov in Russia, and Ernst from Macedonia. Both suffering from malaria, they never wanted to discuss their war experiences with me or anybody else. Ernst enrolled at the TH Stuttgart in special courses for teaching in vocational schools; Carl emigrated to the United States via Ellis Island in 1925.

My secondary school ended in 1924. My father already had arranged for my apprenticeship with a notary in Ulm. My headmaster, however, discussed my interest and good grades in mathematics and science with my parents and tried to persuade them to have me continue my education. My parents finally agreed to the headmaster's suggestion, despite the financial strain associated with it; the hyper-inflation of 1923 had wiped out their savings. I appreciated my family's sacrifice; it increased my continual endeavor to strive for excellence in my education and profession.

I continued my studies at the Ulm *Oberrealschule* (Senior High School). A highlight of my last school year was an excursion to the Deutsches Museum in Munich, one of Europe's famous permanent exhibits for science, technology and industry. I then decided to study engineering; I preferred dealing with the laws of physical science rather than with people.[1] Graduation in 1927 with the *Abitur* (Final Examination) allowed me to enter a university or equivalent institute. For the study of engineering, a six-month factory apprenticeship was mandatory prior to enroll-ment.

---

[1] At that time, that was one of my major misconceptions of the realities of the world. My career taught me a different lesson in this respect.

My brother Ernst occupied a teaching position at a new vocational school in Schweinfurt, the ball bearing city that became famous as an air-raid target in WW II. He offered to let me stay with him and his family. I gratefully accepted and absolved my apprenticeship with the Schweinfurter Präzisions-Kugellager-Werke Fichtel & Sachs A.G. (acronym F&S). Manufacturing included the Torpedo, a patented rear hub for bicycles, and antifriction bearings, both at a production rate of 25,000 per day. As a blue collar worker with black fingernails, I had to perform hard, highly disciplined work, manually, physically and mentally. I learned all metal manufacturing and inspection techniques, scheduling including the recently heralded JIT (Just In Time) deliveries, zero scrap rates, etc. I extended my apprenticeship to 18 months, with the last months in the Design Department. There we established the groundwork for the later-famous Sachs engines by operating, disassembling and analyzing all available domestic and foreign specimens of small engines. My work at F&S provided a good foundation for the engineering profession in general and the techniques for mass-producing precision components and aggregates in particular. Since then, I feel comfortable with both blue and white collar workers. Though I "lost" a year, I still feel it was a worthy, excellent investment for my career.

Enrolling in the TH Stuttgart in the Fall of 1928, I studied electrical and mechanical engineering. The faculty included professors with international reputations. To name a few prominent ones: Carl von Bach (1847–1931, materials and their testing), Richard Grammel (1889–1964, dynamics and mechanical vibrations), Wilhelm Kutta (1867–1944, mathematics, Kutta-Joukowski formula for the lift of airfoils), Erich Regener (1881–1955, physics, cosmic ray investigations with high-flying balloons), and Paul Bonatz (1877–1956, architect; his Stuttgart Central Station of 1913 is still modern today).

In my second year the TH celebrated its 100th anniversary. For the full month of May 1929, faculty and students reigned supreme; we students received free tickets for commuter trains, tram cars, theaters, movie houses, etc. A variety of activities testified to the excellent cooperation of the city and its citizenry with their TH. Highlights were a *Festvorstellung* (gala performance) in the *Staatstheater*, a daylight parade of faculty and students through the inner city as well as a torchlight parade to the *Neues Schloβ*, and last but not least, a *Festkommers* (festival with beer and pretzels) in the *Stadthalle* (sports arena) sponsored by the local breweries. As a member of the A.W.V. Makaria (academic-scientific fraternity) I proudly participated in most of these festive events.

I graduated as Diplom-Ingenieur[2] in 1932, i.e., during the Great Depression, without a job opportunity. The TH, with the financial assistance of the Verein

---

[2] The grade of Diploma Engineer lies between a Bachelor of Science (B.S.) and a Master of Science (M.S.) degree.

Deutscher Ingenieure (VDI, Society of German Engineers), other organizations and the local industry, was conducting a study and research program for graduate engineers. It supported my further education.[3]

After a year in this program, I was fortunate to obtain jobs in Stuttgart and Berlin that provided me with enough time and resources for doctoral work. In 1937, the TH Stuttgart awarded me the Doctor of Engineering with my thesis *Schalldämpfer für Rohrleitungen* (sound dampers for pipelines). The VDI included excerpts in its journals, particularly my concept and test results of a very effective, adjustable Helmholtz resonator for the attenuation of gas vibrations in a pipe over a wide frequency range. It also published and promoted my thesis as a book. It received favorable reviews in 25 domestic and foreign business newspapers, professional journals and magazines, two of them in the U.S. These accolades also helped me to gain a position in the engine industry at Bramo in Berlin-Spandau.

In Berlin I occasionally called on Tante Rickchen (Aunt Fredericka), a native of Jungingen who had attended my parents' wedding. There I met at dinner my future wife M. Magda Pfister who was staying with the daughter of our hostess. It was a fortunate coincidence, two Swabians meeting in the Berlin metropolis. After a one-year courtship, our wedding took place in the Summer of 1938 in the *Friedenskirche* (Peace Church) in Heilbronn am Neckar. We also utilized this stay to survey the industry in southern Germany for a desirable relocation from Berlin. Our honeymoon trip with a DKW cabriolet took us along the romantic Neckar and Rhine valleys to the Harz mountains and to Berlin. I then decided to accept an offer from the Hirth-Motoren Company in Stuttgart-Zuffenhausen as research and development engineer in its Applied Research Department. In December 1938 we settled in a Stuttgart suburb.

After living 14 years in the United States, my brother Carl paid his first visit to Germany, arriving in Stuttgart in July 1939, on the day Magda gave birth to our first baby, Rose-Marie. Her christening gave Carl the happy opportunity to meet Magda's parents and family. On a visit to Jungingen we assembled for a photo of my parents and their children (Figure B.2). This was going to be the last time we were all together. To our question on the length of his stay, Carl surprised us to our disbelief with his remark, "I'll have to return soon because there will be war in Europe!" His hunch proved right; he departed with the S.S. Bremen on her last trip to New York. On September 1, Hitler's invasion of Poland triggered World War II.

Further events of my professional and family life are included in the previous chapters of this book.

---

[3] I am happy to note that SAE is pursuing a similar program for engineering students.

Fig. B.2. The Bentele family, from left: (front row) father Carl Christian (1869–1951), mother Walburga (1869–1944), Wilhelmina (Mina, 1896–1960); (back row) Max (1909–), Margarete (Gretel, 1901–1989), Carl (1897–1968), Barbara (Babe, 1903–1978), Ernst (1899–1970).

Concerning my ancestry and family background, Carl Bentele, an auditor/accountant of the city of Ulm, used his retirement to investigate the history of the Bentele clans. We were unknown to him as he was to us. Assuming there must be more Benteles around, he searched for their origins and whereabouts. It took him a few years and the perusal of scores of documents in secular and church archives. He traced the name back to the Lords of Pentelingen/Bentelingen in the years 1070 to 1095 A.D. Other documents of 1200 A.D cited the names Bendelin and Bändelin. Late 14th century archives abound with the names of Bentelin, Bentelli and Bäntelli, all living in southern Germany, eastern Austria and northern Switzerland. They comprise craftsmen, tradesmen, artists, sculptors, painters, and writers. He established in detail 25 branches and named them mostly according to their predominant locations.

Our own Bentele family belongs to the Günzburg/Leipheim/Ulm branch; our family tree is continuously documented back to the year 1500. It contains mostly craftsmen and farmers, occasionally municipal officials, but no celebrities who made the history books. Our coat-of-arms, dated 1400, depicts on a silver background a bull's head in red. My brother Ernst, an excellent amateur artist, painted it on parchment as his wedding gift to us. He later wrote and illustrated a handbook on calligraphy which was distributed worldwide.

Some noteworthy items of Ulm's history are inscribed in my memory. The *Ulm Münster* (minster) dominates the city's skyline from all directions as an impressive Gothic cathedral. During a period of prosperity in 1377, the Ulm citizenry had laid its foundation stone and ceremoniously showered it with money, indicating the building as a citizen church, not one founded by a bishop or a prince. In 1890 the church was finally completed; the top stone knob was put on the main tower, elevating it to the highest cathedral in the world with 161 meters (528 feet). The suspicion that the spire had been "stretched" in order to exceed the 156-meter high Cologne Dom was dismissed as nonsense. A gothic structure prohibits such a perversion, the architects say. Still, a then-issued illustrated brochure compares the Münster with ten other European cathedrals, from St. Peter in Rome to St. Paul in London and St. Giraldi in Seville to the Stephansdom in Vienna, proudly depicting the Ulm Münster as the highest. As a remarkable sight in appearance and height, it attracted visitors from all over the world, and still does.

The *Münsterfest* of June 1890 followed the official inauguration of this magnificent monument. It included a parade in which my father proudly participated with the Jungingen contingent wearing traditional costumes. As youngsters we often climbed all steps to the top, the octagon deck, to demonstrate our athletic prowess and to enjoy the grand view of the town and country, our small village four miles to the north, and on a clear day the majestic Alp mountains 90 miles to the south. This tradition continues with our children and grandchildren.

The 100th anniversary of the completion of the Ulm Münster was celebrated in May 1990. The Lord Mayor of Ulm greeted the ecclesiastical and secular official guests in the *Rathaus* (City Hall), again emphasizing the Münster as a citizen church. Highlights of the anniversary celebration were: a *Ständchen* (serenade) to the church by several choral societies, one from East Germany, and joined by other attendees; an illumination of the Münster with superimposed pictures of past events such as the half-finished building, the flight of the Tailor of Ulm, among others; and a concert presented by 10,000 trombone players, another record in numbers.

Schwörmontag (oath-Monday) is a unique Ulm holiday celebrated each August and dating back 400 years. From an alcove of the *Schwörhaus*, the Lord Mayor delivers in the morning to the assembled citizenry his state-of-the-city message, ending with the solemn oath that he will be impartial to rich and poor alike, without reservation. The official ceremony is followed by traditional festive events on the Danube river and in the streets and squares of the city. Locals in antique garb perform age-old competitions; citizens also take part in parades, dances and candle serenades. Thousands of spectators and visitors rejoice in the serene, jubilant and romantic events.

Ulm is also one of the cradles of aviation. As schoolchildren we used to sing in Swabian dialect, "D'r Schneider von Ulm hot's Fliege probiert, no hot en d'r Teufel en d'Donau neig'führt!" (the tailor of Ulm tried to fly, but the devil dumped him into the Danube river). The tailor, Albrecht Ludwig Berblinger (1770-1829), had glided with his self-built flying machine (Figure B.3) in a vineyard from one cabin to a lower one; he also had flown from Ulm mountainsides. To impress the King of Württemberg on his visit to Ulm in May 1811, Berblinger was wheedled to perform a spectacular glide across the Danube, from the *Adler-Bastei* (eagle bastion), a tower of the city wall, to the Bavarian river bank, a distance of 40 meters (131 feet). Analyzing his previous achievements, his calculations required an additional wooden structure. With this he saw a chance for success, and he agreed to the performance. His flight failed, however, most probably due to downdrafts and ill winds. He ended in the river, physically unhurt, but suffering the jeers of the spectators and the continuous scorn of the Ulm populace. He was unable to resume his tailor business. "Who would order a coat from Berblinger; one would have to fear it could have wings instead of sleeves!" was the common saying. After a second, troubled marriage, he died as a pauper. Eighty years later, Otto Lilienthal (1848-1896) was more successful with his gliders, but lost his life in a flight.

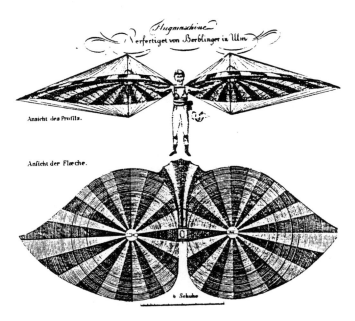

Fig. B.3. Albrecht Ludwig Berblinger's flying machine (1811).

In 1986, to commemorate and probably exonerate its aviation pioneer, the City of Ulm inaugurated the 175th anniversary of Berblinger's venture with an International Competition offering a cash prize of 50,000 Deutschmarks (approximately $25,000). Only gliders similar to that of the tailor were permitted. All but one of the two-dozen final competitors shared the tailor's fate by falling into the river. With a plane constructed by his father, a 19-year-old made it over the Danube, to the delight and applause of officials and spectators. The tailor's ridicule was thus transformed to admiration.

The University City Ulm represents the best of the Swabian enterprising spirit, diligence, and perseverance. As a unique high-tech center it pursues research in medicine and electronics, manufacturing of top firefighting equipment and vehicles, luxury tour buses, specialty trucks, diesel engines, and machine tools, among others. "In Ulm, you can see the future," said a prominent Federal official. Its sons of the last century, Max von Eyth (1836-1906), engineer in agriculture and author of tales from the industrial world, and Albert Einstein (1879-1955), would be proud of its present environment, culture and achievements. They probably would not have had the urge to leave for other countries. Similar feelings apply to myself. If the present conditions had been prevalent when I was at the start of my career, I might have stayed "at home in Swabia." In that case, my *Wanderlust* would have presented a nagging demand, I would hardly have had such a fascinating and rich life, made so many friends worldwide, and gathered experiences in all respects. If this book would have been written at all, it certainly would be different.

Index

A/S Vapenfabrikk Kongsberg KG 2/3, Radial-inflow turbine, 49
AAPS. *See* Alternative Automotive Power Systems
AB Volvo, 228
Abingdon-Cross engine, 11
Acoustics, Helmholtz resonator sound dampers, 142
Adiabatic diesel engine, 226
Adjustable Helmholtz resonator, 258
Advanced gas turbine engines, 189
Advanced Integrated Propulsion System (AIPS), 227
AEG. *See* Allgemeine Elektrizitäts-Gesellschaft
Aero-engines, turbocharging in conjuction with engine-driven supercharger, 15
Aeroacoustics, 142
Aerodynamics, superchargers, 13
Aerodynamische Versuchsanstalt (AVA), 45, 52, 83-84, 91, 102
Aeroelastic instability, gas turbine compressor, 148-149
Aeroelastic stability, compressor blades, 171
AGT 1500, tank gas turbine, 191-193, 197
AIAA. *See* American Institute for Aeronautics and Astronautics
Air Force Aerospace Research Laboratories, 65
Air Force Museum, 61, 66
Air Ministry, 21, 25-27, 29, 34-35, 38, 45, 51, 61, 68, 82, 116
    Conference on cooled exhaust turbines, 25
    jet engine development program, 44
Air Research Division, Garrett Corporation, 82
Air-cooled aero-engines, Hirth Motoren Company, 7
Air-cooled blade, turbine cooling, 146
Air-cooled DISC engine, Curtiss-Wright, 211
Air-cooled engines, 6
Air-cooled gasoline engine, Curtiss-Wright, 211
Air-cooled jet engines, 146
Air-cooled radial turbocharger, BMW 801 J engine, 26
Air-cooled turbines, 26, 57-61, 81, 100, 146
    ceramic rotor, 121
Air-cooled turbochargers, 146
Air-cooling, Wankel aircraft engines, 211
Air-cushion vehicles,
    alternative railroad, 144
    Curtiss-Wright, 144-145
Aircraft, Curtiss-Wright X-19 experimental aircraft, 178
Aircraft engines,
    BMW 003 single-shaft jet engine, 47
    compressor blade failures, 189-191
    propulsion, 33

TL Turbinen-Luftstrahltriebwerk, 61
see also specific engines by name
Aircraft propulsion,
    engine/fan jets, 33
    gas turbines, 91, 187, 221
    ML Motorrückstößer-Luftstrahltriebwerk engine jet propulsor, 61
    PTL Propeller-TL turboprop, 61
    ramjet, 33
    turbojet, 33, 61
    ZTL Zweikreis-TL turbofan, 61
Airflow pattern, Wankel diesel application, 166
ALF 502 high-bypass turbofan, British Aerospace B Ae 146, 201
Allgemeine Elektrizitäts-Gesellschaft (AEG), 49
Allison Division, General Motors, 176-177
Allison regenerative turboprop engine, 176-177
Alternative automobile engines, Daimler-Benz, 225
Alternative Automotive Power Systems (AAPS),
    EPA, 216
    Symposium, 212
Alternative engines, 189, 208, 216, 223, 225
Alternative fuels, 218
    Low Pollution Power Systems Development Program (LPPSD), 216
Alternative railroad,
    air-cushion vehicles, 144
    magnetic levitation (MAGLEV), 144
American Institute for Aeronautics and Astronautics, 68, 243
    Symposium for the 50th Anniversary of the first jet-powered flight, 68
American Institute of Aeronautics and Astronautics (AIAA), 68, 165, 243
American Society of Mechanical Engineers, 22
"Anglo-American Jewel" J65 jet engine, Curtiss-Wright, 161
Applied Research Department, Hirth Motoren Company, 7-8
Arado 79, Hirth engine, 7
Arado Ar 234 reconnaissance aircraft, Junkers Jumo 004 engine, 47
Armstrong-Siddeley, 118, 146, 148, 161
Armstrong-Siddeley engine, 137
Army Aviation Systems Command (AVSCOM), 189-190
Aspin, rotary valve engine, 10
Astronaut "cap pistol", WAD aviation and aerospace design studies, 177
ATAR. *See* Atelier Aeronautique de Rickenbach
ATAR engine, SNECMA, 122
ATD. *See* Automotive Technology Department
Atélier Aéronautique de Rickenbach (ATAR), 83
Austin Co. Hayes transmission, toroidal traction drive, 177
Automatic transmission, Heinkel touring scooter, 112
Automobile engines,
    alternative engines, 189, 208, 216, 223, 225

alternative fuels for, 218
gas turbine technology, 88
Automotive engine technology, 215
Automotive gas turbine,
Ford, 225
Mahle piston factory, 120-122
Automotive gas turbine engines, 83-84, 120-121, 194-195, 220, 222, 225
Automotive gas turbines, ceramic components, 223
Automotive History Collection, Detroit Public Library, 241-242
Automotive Technology Department (ATD), 216,225
AVA. *See* Aerodynamische Versuchsanstalt
AVCO,
Lycoming Stratford Division, 82, 103, 178, 182, 187-203, 206
Pollution Conference, 194
Aviation and aerospace design studies, WAD, 177
AVRO, Vulcan delta-wing bomber, 118
AVSCOM. *See* Army Aviation Systems Command
Axial turbine, 68, 195
Axial-flow,
jet engine, 41-48
supercharger, 13
turbomachinery, 67
He S 30 engine, 41-44
Axial-flow compressors, 67-68, 81, 84, 200
Ayres, Robert U., 218

Baer, rotary valve engine, 10
Baker, R.J.S., 11
Barber, Everett M., Texaco Combustion Process (TCP) stratified charge, 166
BASF, Technische Prüfstand (Engine Laboratory), 29
BASF Otto-Diesel aircraft engine, Wilke, Wilhelm, 29
Bayerische Motorenwerke (BMW), 26
Beier, 27
Beier CVT, 181
Beier transmission, 112
Heinkel touring scooter automatic transmission, 112
Toroidal traction drive, 177
Beier variable-speed drive, 27
Bensinger, Daimler-Benz Wankel engine, 8
Bensinger flat-disk valve engine, Daimler-Benz DB, 8
Bensinger, Wolf-Dieter, 8
Bentele, Brigitte Regina, 46
Bentele, Magda, 46, 69-72, 203, 206-207
Berblinger, Albrecht Ludwig, 261-262
Berner, Roland T., 161, 172, 205
Betz, Albert, 84

Blade vibration, compressor, 148, 171
Blanton, Ray, 207
BMW 003 engine, vibration fatigue turbine blade failures, 47-48
BMW 801 J engine, air-cooled radial turbocharger, 26
BMW (Bayerische Motorenwerke), 10, 21, 26, 33, 38, 47-48, 54, 61, 83
    Bramo, 47
    rotary valve engine, 10
Boeing, 120, 173, 225
Boeing B-17 bombers, 46
    Moss turbochargers, 27, 61
Boeing gas turbine, Kenworth truck, 120
Boeing Stratocruiser passenger airplane, 102
Boeing truck, vehicular gas turbine, 225
Bonatz, Paul, 257
Bosch, "R" engine fuel injection system, 29-30
Bosch Company, 29, 123
Bouchard, Phillipe O., 65
Bowden, Andrew T., 83, 88, 99, 102
Brabazon, Lord, 101
Brabazon passenger airplane, Bristol, 101
Bradley, James J., 241
Braig, Hans, WMF Topfschaufel (tubular or bootstrap blade), 24
Bramo, 1, 47, 258
    BMW, 47
    Siemens Group, 1
Bramo SH 14a engine, FW-61 heliocopter, 1-2
Brandenburgische Motorenwerke (Bramo), Instrumentation and Measurement
    Laboratory, 1
Braun Company, 206-207
Braun, Hans, 206
Braun, Karl Otto, 206
Brayton cycle, gas turbine engine, 222
Brayton cycle (gas turbine), Low Pollution Power Systems Development
    Program (LPPSD), 216
Brenneke, Arthur, 159
Bridge collapse,
    Ohio River Bridge, 23
    Tacoma Narrows suspension bridge, 22-23
Briggs & Stratton, 83
Bristol Aeroplane Company, 3
Bristol Brabazon passenger airplane, 101
Bristol Centaurus, air-cooled engine, 6
Bristol engine, 3, 5-6
    Fedden, Alfred H.R. (Sir Roy), 3-4
    single sleeve aero-engine, 3-4
Bristol Hercules, air-cooled engine, 6

British Whittle Group, 81
Brooklyn Bridge, Roebling, John A., 23
Bruckner, Canis & Co., 52
Burt and McCollum, 3
    single-sleeve engine, 3, 5
Butter Brothers, Sprengniete (explosive rivet), 116
Butter, Karl, 116
Butter, Otto, 116

C.A. Parsons & Co., 83, 88
    coal burning gas turbine, 99-100
    gas turbine design department, 88-89
    gas turbine laboratory, 88, 99
    Marine Division, 88
Cabin scooter "Kabine", Heinkel, 115
Cal Tech, Jet Propulsion Lab (JPL), 220
Campbell diagram, 21
Canadair CL84 tilt-wing aircraft, Curtiss-Wright hollow propeller blades, 178
Canberra twin-jet, Armstrong-Siddeley, 118
Caproni-Campini engine, 64
Carl Zeiss, 60
Carter, A.T.D.S., 148
Centaurus engine, 6
Centrifugal compressor, 198
    tank gas turbine, 89
Centurion tank, 90, 99
Ceramic components,
    automotive gas turbines, 223
    gas turbines, 59-61, 223
Ceramic gas turbine, 222-226
    Ford, 222
    Westinghouse, 222
Ceramic materials, 222, 225
    gas turbine engine, 222
Ceramic turbine "Fernwagen", Daimler-Benz, alternative automobile
  engines, 225
Champion Spark Plug Company, 201, 217
    "Ignition and Engine Performance Conference", 217
Chilton Book Company, 241-242
Chrysler Corporation, 192, 194, 221-223, 225-226
    gas turbine-driven passenger car, 221-223, 225
Chrysler/Lycoming, AGT 1500 tank gas turbine, 192, 226-228
Club der Luftfahrt von Deutschland, 72
Coal burning gas turbine, 100
    C.A. Parsons, 99-100
    fluidized beds under pressure, 100

General Motors, 100
industrial applications, 100
locomotives, 100
steam boilers, 100
Combustion, 2nd Annual Rotating Combustion Conference, 179-181
Combustion system, Joseph Lucas Co., 90
Combustor development, He S 011 jet engine, 57
Commercial applications, stratified-charge Wankel engine, 211
Committee on the Challenge of Modern Society (CCMS), NATO, 216
Committee of the National Academy of Science, 38
Committee to assess compressor disk failures, 189-191
Commuter car, flywheel, 232
Composite hollow propeller blades, 178
    Curtiss-Wright, 178
    Rolls-Royce, 178
Compression ratio, Curtiss-Wright Wankel rotary development, 165-166
Compressor, 198
    advanced gas turbine engines, 148
    blade vibration, 148
Compressor blade failure, T53-L-13 engine, 189-191
Compressor blades,
    aeroelastic stability, 171
    friction dampers, 171
    mechanical dampers, 171
    vibration fatigue, 171-172
Compressor development, He S 011 jet engine, 52-56
Compressors, axial-flow compressors, 67-68, 81, 84, 200
CONCO Medical Company, 206-208
Conference on air-cooling of gas turbines, NACA Lewis Flight-Propulsion
    Laboratory, 146
Conference on Ceramic Gas Turbine Components, Lilienthal Society, 60-61
Conference on cooled exhaust turbines, Air Ministry, 25
Conference marking the 40th Anniversary of the first jet flight, National Air
    and Space Museum, 68
Conference on radial and axial compressors, Royal Aeronautical Society, 68
Conference, Second Annual Rotating Combustion, Curtiss-Wright Wankel
    rotary development, 179
Conference on SST, FAA, 173
Conference in Toronto, Parsons powerplant failure, 101
Conference with Whittle and von Ohain on development of jet engine,
    Air Force Museum, 65-66
Constellation passenger airplane, Lockheed, 102
Continental Motor Corporation, 5
Continually variable transmission (CVT), Heinkel touring scooter, 112
Contractors Coordination Meetings (CCM), 217
Convection cooling, turbine cooling, 146-149

Cooled turbine technology, 148
Cooled turbines, advanced gas turbine engines, 148
Cooling problem, Wankel rotary engine development, 151
Crawford Auto-Aviation Museum, 61
"Cross cool combustion chamber", rotary valve engine, 10
Cross, Roland and Michael, 11
Cross rotary valve, 11
Cross-Baker engine, 11
Cummins Engine Co., 226, 228
   oil-cooled turbocharged diesel, 227-228
Curtiss-Wright, 6, 27-28, 103, 123-125, 151-154, 158-161, 163, 165, 168, 173,
   175, 181-182, 197, 201, 205-206, 209-211, 228, 231, 241-243
   air-cooled DISC engine, 211
   air-cooled gasoline engine, 211
   air-cushion vehicles, 144-145
   "Anglo-American Jewel" J65 jet engine, 161
   Composite hollow propeller blades, 178
   Cyclone 9 air-cooled piston engine, 137
   DISC experimental engine, 168
   hollow propeller blades,
      Canadair CL84 tilt-wing aircraft, 178
      Curtiss-Wright X-2 experimental aircraft, 178
   Research Division, Muffling Turbo Compound TC 18, 142
   rotary engine technology, 182
   Sapphire J 65 turbojet engine, 137
   transpiration cooling of turbine blades, 147-148
   Turbo Compound TC 18, 6, 27-28, 197
   Turbo Compound TC 18 supercharger, 27
   Twin-spool jet and ramjet engine, 148
   U.S. Patent Wankel engine, 153-154
   Vertical takeoff and landing airplanes (VTOL), Wankel engine, 159
   Wankel rotary engine development, 151, 153, 156, 158
      2nd Annual Rotating Combustion Conference, 179-181
      advancing the engine, 164
      axial flow cooling, 157
      burning heavy fuels, 164
      champagne luncheon, 160
      compression ratio, 165-166
      design parameters, 164, 179
      DISC (direct injection stratified charge), 166-168
      engine applications, 159
      engine geometrics, 158, 179
      gas sealing problem, 159, 164
         Perfect Circle Company, 159
      gas sealing system, 159, 179
      low compression ratio, stratified-charge, 166-168

multifuel engine, 164, 179
presentation at SAE, 164-165
Princeton University meeting, 165
RC1-60 experimental engine, 161
RC4-60 experimental engine, 161
rotor and housing structure and cooling, 157, 179
sleeve bearings for rotor and power shaft, 157
use of heavier fuels, 165-168
Wankel rotary engine applications,
  boats, 159
  buses, 159
  cars, 159
  generator sets, 159
  industrial drives, 159
  tractors, 159
  trucks, 159
Wright Aeronautical Division (WAD), 81, 137
X-19 experimental aircraft, 159, 178
  Vertol/Bell X-22 Osprey, 178
X-2 experimental aircraft, Curtiss-Wright hollow propeller blades, 178
Cyclone 9 engine, Curtiss-Wright, 137
Cyclone aircraft engine, Wright, 123

D. Napier & Son Ltd., 27
DAC. *See* Douglas Aircraft Company
DAC/CW, "Project Smoothie", 142
Daimler Motoren-Gesellschaft (DMG), Knight double-sleeve engine, 3
Daimler-Benz, 8, 33, 38, 47, 63, 83, 108, 123, 139, 141, 194, 201, 225
Alternative automobile engines ceramic turbine "Fernwagen", 225
DB 600, flat-disk aero-engine, 8
DB 612, Bensinger's flat-disk valve engine, 8
Mercedes 300, 141
Wankel engine, Bensinger, 8
ZTL DB 007 experimental engine, 63
Daimler-Benz/Heinkel-Hirth, PTL DB/He S 021 experimental engine, 63
Dart Aircraft Company, 118
De Havilland, DH 110 fighter, 118
De Havilland Company, 68, 90, 118-119
radial compressor, 90
de Saunier, L. Baudry, 230
Decher, Siegfried H., 198, 201
deLavaud, Sensaud, 154
Department of Energy (DOE), 216
Detroit Public Library, National Automotive History Collection, 212, 241
Deutsche Akademie (German Academy), 2
Deutsche Versuchsanstalt für Luftfahrt (DVL), 8, 13, 16, 19, 21, 26, 54, 63, 82

Institut für Motorische Arbeitsverfahren und Thermodynamik, 24
Institut für Strömungsmaschinen (Institute for Turbomachinery), 14
Deutsches Museum, Munich, 160
Deutz Company, 181
Development of tank gas turbine, C.A. Parsons & Co., 88-99
DGLR, Symposium "50 Jahre Turbostrahlflug - 50 Years of Jet-Powered
  Flight", 69
DGLR (Deutsche Gesellschaft für Luft und Raumfahrt), 69, 72, 139
DH 110 fighter, De Havilland, 118-119
Diesel, high compression ratio, 168
Diesel engine, 120, 160, 181, 195, 197, 215, 218, 227-229
  turbocharger development, 83
Diesel passenger car, General Motors, 141
Diesel, Rudolf, 160, 166, 181, 216
Diesel tank engine, 226-228
Diesel-driven Leopard II tank, 227
Diesel-driven vehicles, 230
Differential gas turbine, 198-200
Differential turbine application,
  helicopter, 198
  turboprop, 198
  vehicle, 198
Differential turbine principle, Kronogard Turbine Transmission, 225
"Diluent stator", WAD aviation and aerospace design studies, 177
DISC (direct injection stratified charge) engine, Wankel rotary development,
  166-168, 228
DISC experimental engine,
  Curtiss-Wright, 168
  wide range of fuels, 168
DISC multifuel stratified-charge engine, 210
  FVRDE, 210
  jet fuel, 210
DKM (Drehkolbenmaschine dual rotation machine), NSU/Wankel
  project, 152-153
DKW (Das Kleine Wunder) cabriolet, 4, 258
DMG. *See* Daimler Motoren-Gesellschaft
DOE. *See* Department of Energy
Dornier, Claude, 34, 118
Douglas Aircraft Company (DAC), 141
Drives, toroidal traction drive, 177
Dual-cycle engine, jet and ramjet, 123
DuPont Company, 116
Dutch Philips Company,
  Stirling engine, 219-220
  Stirling engine-driven bus, 220

DVL. *See* Deutsche Versuchsantalt fur Luftfahrt
DVL turbocharger, 13-14, 16
DVL/Hirth, 59
DVL/Hirth turbocharger, 16, 45, 59
    Junkers supercharged engine Jumo 211, 16
    KHD Dz 710 diesel engine, 26
    model 9-2216, 16-19
    model 9-2281, 16-21
    model 9-2426, Junkers Jumo 222 engine, 26
    vibration fatigue turbine blade failures, 16-21
Dynamit-Nobel-Konzern, 116

Eagle engine, Rolls-Royce, 6
Early jet engine development, 65-72
Ebert drive, Hydrostatic axial piston pump transmission, NSU scooter, 112-113
Ebert, Heinrich, Hydrostatic axial piston pump transmission, 112-113
Einstein, Albert, 262
Electric bus, 232
Electric car, Whisper electromobile, 231
Electric cars, 229-231
    Low Pollution Power Systems Development Program (LPPSD), 216
    Sinclair C5, 231
Electric motorcycle, 229
Electric vehicle, 229-231
Encke, 52, 55
Energy Research & Development Administration (ERDA), 216
Engine applications, Curtiss-Wright Wankel rotary engine development, 159
Engine development, DISC (direct injection stratified charge), 166-168
Engine geometry, rotary engine, 187
Engine/fan jets, aircraft propulsion, 33
Engines,
    advanced gas turbine engine compressors, 148
    air-cooled DISC engine, 211
    air-cooled engines, 6
    "Anglo-American Jewel" J65 jet engine, 161
    axial flow cooling, 157
    axial-flow jet engine, 41-47
    BMW 003 engine, 47
    burning heavy fuels, 164
    compression ratio, 165-166
    compressor blade failure, 175, 189-191
    compressor blades, 148, 171-172, 189-191
    Cyclone 9 air-cooled piston engine, 137
    design parameters, 164, 179
    development and engine applications, 159-177
    DISC experimental engine, 168

engine geometries, 158, 179
jet engine, 65-66
reciprocating engines, 142
rotary engine technology, 182
Sapphire J 65 turbojet engine, 137
Turbo Compound TC 18 engine, 6
Twin-spool jet and ramjet engine, 148
U.S. Patent Wankel engine, 153-154
Wankel engine development, 151, 153, 156
Engines *see also* Curtiss-Wright
Environmental Protection Agency (EPA), 216
　　Alternative Automotive Power Systems (AAPS), 216
EPA. *See* Environmental Protection Agency
EPA/NATO, Low Pollution Power Systems Development Program, 216
ERDA (Energy Research & Development Administration), 216
Esso Oil Company, 11
Esso-Cross engine, 11
Exhaust gas turbine, turbocharger, 13

FAA. *See* Federial Aviation Administration
Faltschaufel, turbine blade, 24-25, 57
Farnborough Flying Display and Exhibition, Society of British Aircraft
　　Constructors, 117
Farrar, Earl V., 81, 123, 137
Fedden, Alfred H.R. (Sir Roy), 3-4, 7
　　Bristol engine, 3-4
Federal Aviation Administration (FAA), 172-173
　　Conference on SST, 173
Fiat, 194
Fichtel & Sachs, 161
　　Wankel engine license, 161
Fichtel & Sachs (F&S), Schweinfurter Präzisions-Kugellager-Werke, 257
Fighting Vehicles Research and Design Establishment (FVRDE), 88-90, 93,
　　99, 102, 210
　　DISC multifuel stratified-charge engine, 210
　　MoS, 88
　　tank gas turbine, 210
First jet fighter, He, 38, 280
First jet propelled flight,
　　He 178 aircraft with He S 3 B engine, 35-41
　　Heinkel experimental airplane, 35-41
　　von Ohain, Hans, He S 3 B engine, 35-41
Flat-disk aero-engine, Daimler-Benz DB 600 , 8
Flat-disk valve engine, 7-10
Flexible compressor blades, 171-175
Flexible compressor blades/friction dampers, WAD, 175

Flying Display and Exhibition of the Society of British Aircraft
  Constructors, 101
Flywheel,
  commuter car, 232
  subway train, 232
Focke, Henrich, 1, 122
  FW-61 helicopter, 1
Fokker, Anthony, 122
Ford Motor Company, 194, 220, 222-223, 225, 231
  automobile gas turbine, 225
  ceramic materials, 225
  ceramic turbine, 222
  model, 194, 704
  Stirling engine, 220, 225
Franz, Anselm, 44, 68, 82, 188
French Air Force, 80
French Hyperbar engine, 28
Friction dampers, compressor blades vibration fatigue, 171-172
Friedrich, Rudolf, 41, 52
Froede, Walter, 152, 164
Fuels, alternative fuels, 218
FVRDE. *See* Fighting Vehicles Research and Design Establishment
FW-61 helicopter, Focke, Henrich, 1
  Bramo SH 14a engine, 1-2

Garrett Corporation, 49, 228
  Air Research Division, 82
Garrett/Ford, 223
Gas generator, 89, 94, 99, 192
Gas generator rotor, 91
Gas sealing problem, Wankel rotary engine development, 151, 159
Gas sealing system, Curtiss-Wright Wankel rotary development, 159
Gas turbine, 179
  automotive application, 120
  ceramic materials, 223
  Chrysler, 223
  commercial vehicle application, 228-229
  industrial application, 99-103
  Ljungstrom rotary regenerator, 91
Gas turbine compressor,
  aeroelastic instability, 148-149
  vibration fatigue blade failure, 148
Gas turbine design department, C.A. Parsons & Co., 88
Gas turbine development, 82, 101, 120
Gas turbine development for industrial and marine applications, Parsons, 102

Gas turbine engine, 6, 59, 82-84, 87, 90, 94, 96, 98-99, 103, 121, 168, 187, 194, 197, 215-216, 220-223, 227-228
    Brayton cycle, 222
    ceramic components, 59-61, 223
    ceramic materials, 222
    components, 195
    for helicopters, 82
    for ship propulsion, C.A. Parsons & Co. Marine Division, 88
    industrial application, coal burning, 99
    laboratory, C.A. Parsons & Co., 88, 99
    materials, thermal shock tests, 98
    powerplant, 99
    technology, 205
        C.A. Parsons & Co., 88
        industrial gas turbine, 88
        steam turbine, 88
        tank gas turbine, 88
        vehicle gas turbine, 88
    turboshaft engine, 197
Gas turbine-driven passenger car, 221, 229
    Chrysler Corporation, 221-222
    Rover Jet 1, 93, 120, 221
Gas turbine-powered automobile, 225
Gas vibrations in pipelines, 1, 258
Gasahol, alternative fuel, 218
Gear-driven supercharger, 13-14
General Dynamics, 173
General Electric Company (GE), 21, 41, 148, 173-176, 192, 226, 228
    GE-4 SST engine, 174
    J79 engine supersonic fighter/bombers, 174
    J93 B-70 bomber, 174
    single-spool jet engine, 148
General Electric/Ryan, vertical takeoff and landing aircraft (VTOL), 176
General Motors (GM), 100, 141, 176, 192, 194, 209-210, 220, 222, 231
    Allison Division, 176, 223
    coal burning gas turbine, 100
    diesel passenger car, 141
    gas turbine, Greyhound Super 7 Turbocruiser bus, 222
    metallic gas turbines for bus, 222
    metallic gas turbines for industrial applications, 222
    metallic gas turbines for truck, 222
    NSU/Wankel license, 209
    Stirling engine, 220
    Teledyne Continental diesel engine, 192
    Wankel engine, 220
    Wankel project, 209-210

German Ministry of Defense, 173
German National Prize for Art and Science awarded, 1938, Heinkel, Ernst, 35
Gibb, Sir Claude D., 88, 96, 99, 101
Goetze, A.G., 179
Grammel, Richard, 257
Greyhound Super 7 Turbocruiser bus, General Motors gas turbine, 222
Gulf and Western, 231
Günter, Siegfried, 82, 122, 178
    Russian MIG fighters, 122

H-Type Sabre engine, Napier Company, 6
Hahn, Max, 35-36
Hahn, Otto, 84
Hanaway, William L., 152-153
Hawthorne, E.P., 90, 96
Haxel, Otto, 84
Hayes transmission, Austin Co., 177
He 70, 82
He 111 bombers, Heinkel, 118
He 177 bomber, 35
He 178 aircraft with He S 3 B engine, first jet propelled flight, 35-41
He 280 first jet fighter, He S 8 A engine, 38
He 280 twin jet fighter, jet engine He S 8 A, 34
He S 011, 122
    compressors, 83
    jet engine, 59, 61
        combustor development, 57
        compressor development, 52-56
        Heinkel-Hirth, 50-56
        mixed-flow wheels, 52, 54
        turbine development, 57-61
        vibration fatigue turbine blade failure, 54, 57-61
He S 053 jet engine, Heinkel, 122-123
He S 30 engine, 48
    axial-flow turbomachines, 41-44
    Müller, Max Adolf, 41-44
    Wagner-Müller-Heinkel, 41-44
He S 3 B engine, liquid-fuel combustion, 38
He S 8 A engine, 48
    He 280 first jet fighter, 38
    vibration fatigue turbine blade failure, 49
    von Ohain, 48
Heat exchanger, 194, 228
    liquid metal regenerator, 175-176
Heinkel, 34, 44, 49-51, 64, 82, 105-106, 108-110, 113, 115-119, 122-123,
    137, 146

aviation and space ventures, 106
cabin scooter "Kabine", 115
engineering development, 106
gas turbine, 106
He 111 bombers, 118
He 70, 82
He S 011, 122
He S 053 jet engine, 122-123
Hilfsmotor bicycle auxiliary engine, 113
jet engine, 34-44
jet engine compressor development, 50-56
Kasino, 122
OEM engines, 106-108
rocket propelled He 117, 176
Sonderentwicklung Department (Special Development), 36
touring scooter, 108-116
    automatic transmission, 112
    Beier transmission, 112
    continually variable transmission (CVT), 112
    Ebert Hydrostatic axial piston pump transmission, 112-113
    "Heinkel Tourist", 108-116
Heinkel, Ernst, 34-38, 41, 48, 102, 105-106, 109, 111, 113-119, 122-124, 181
    German National Prize for Art and Science awarded, 1938, 35
    Hirth Motoren acquisition, 34
Heinkel experimental airplane, First jet propelled flight, 35-41
Heinkel Group, Maschinenfabrik G.F. Grotz, 80
Heinkel He 70, Siegfried, Günter, 82
Heinkel He 111 aircraft, 16
Heinkel and Hirth, 71, 208
Heinkel jet engine He S 011, Kombinationsverdichter, 50
Heinkel, Karl Ernst, 106, 117
Heinkel, Lisa, 106, 117
Heinkel OEM engine,
    Schnürle loop-scavenging scheme, 106
    Tempo Matador, 106-108
Heinkel-Butter, Sprengniete (explosive rivet), 116
Heinkel-Hirth, 24, 49, 61, 63, 77-78, 80-84, 105, 120, 122, 146, 478
    He S 011 jet engine, 50-56
    He 280/He S 8 A, 50
    jet engine development, Wolff, Harald, 48
    ML He S 50 experimental engine, 63
    Schif, Curt, 48
Heinkel-Hirth/Daimler Benz, PTL DB/He S 021 experimental engine, 63
Heisenberg, Werner, 84
Helmholtz resonator, 142, 258

Helmholtz resonator sound dampers,
    Aeroacoustics, 142
    Muffling Turbo Compound TC 18, 142
Hentrich, Paul, 88, 92
Hercules engine, 6
Hesselman engine, stratified-charge, 166
Hilfsmotor bicycle auxiliary engine, Heinkel, 113
Hindenberg syndrome, Hydrogen fueled vehicles, 218
Hirth, Albert, 7
Hirth engine, Arado, 7, 79
Hirth engines, 7
Hirth and Heinkel, 71, 208
Hirth, Hellmuth, 7, 35, 120
Hirth Motoren acquisition, Heinkel, Ernst, 34
Hirth Motoren Company, 7, 14-16, 18, 21, 23, 29, 33-34, 38, 50-51, 59,
    120, 258
    air-cooled aero-engines, 7
    Applied Research Department, 7-8
    Otto-Diesel type "R" engine, 29
    turbocharger, 14-15
    turbochargers for reciprocating aircraft engines, 33-34
Hirth "R" engine program, 29-31
Hoeppner, Ernst, 152
Hollow air-cooled blades, turbine cooling technology, 146
Hollow propeller blades, Curtiss-Wright, 178
Hryniszak, Waldemar, 88, 92
Hurley, Roy T., 123, 142, 144, 151-152, 156, 158-161, 163, 181-182
Hydrogen fueled vehicles,
    Hindenberg syndrome, 218
    Stevens Insitute of Technology study, 218
Hydrostatic axial piston pump transmission,
    Ebert, Heinrich, 112-113
    Heinkel touring scooter, 112-113
Hyperbar engine, 28

"Ignition and Engine Performance Conference", Champion Spark Plug
    Company, 217
Industrial applications,
    Beier transmission, 112
    Beier transmission, Toroidal traction drive, 177
    coal burning gas turbine, 100
    gas turbine, 99-103, 222
Industrial gas turbine, gas turbine technology, 88
Industrial turbines, transpiration-cooling, 175
Industrial Turbines International (ITI), 228

Institut für Motorische Arbeitsverfahren und Thermodynamik, DVL, 24
Institut für Strömungsmaschinen, DVL, 14
Internal combustion engine, 1, 222
Internal water-cooling of gas turbines, hollow blades, 100
Isuzu, 231
ITI, 228-229
ITI-500, gas turbine engine, 229
ITI-GT601, gas turbine engine, 228

J65 engine, turbine cooling technology, 146-147
J65 jet engine, 148
J79 engine supersonic fighter/bombers, GE, 174
J93 B-70 bomber, GE, 174
Jet engine, 6, 25, 34, 64, 66, 94, 96, 120, 123, 181, 215
    air-cooling, 146
    compressor development, Heinkel, 50-56
    development, 59, 101
    development program, Air Ministry Research Division, 44
    He S 8 A, He 280 twin jet fighter, 34
    Heinkel, 34-44
    technology, 205, 208
    Whittle, 28
Jet fuel, DISC multifuel stratified-charge engine, 210
Jet plane He 178, Siegfried, Gunter, 82
Jet Propulsion Lab (JPL), Cal Tech, 220, 223
Jet and ramjet, dual-cycle engine, 123
John Deere Company, 165, 210
    multifuel stratified-charge engines, 210
    North American Wankel rights, 210
    SCORE (Stratified Charge Omnivorous Rotary Engines), 210
    twin-rotor turbocharged stratified-charge aircraft engine, 210
Jones, Charles (Charlie), 157, 210
    stratified-charge rotary engine, 210
Joseph Lucas Co., combustion system, 90
Junkers Aircraft Engine Division, 38, 44
Junkers Company, 16, 21, 26, 33, 38, 41, 44-46, 49, 54, 82-83
Junkers Ju 287 experimental bomber, Junkers Jumo 004 engine, 47
Junkers Jumo 004, axial-flow jet engine, 44-47
    Arado Ar 234 reconnaissance aircraft, 47
    Junkers Ju 287 experimental bomber, 47
    Messerschmitt Me 262 twin-jet fighter, 47
Junkers Jumo 004A engine, Messerschmitt Me 262 fighter plane, 45
Junkers Jumo 004B, vibration fatigue turbine blade failures, 45-46, 201
Junkers Jumo 211 , supercharged engine, 16
Junkers Jumo 222 engine, 26
    DVL/Hirth turbocharger model 9-2426, 26

K-Cycle engine, 214-215
Kamm, Wunibald, 63
Karman Vortex Streets, 23
Karol engine, "Split Cycle Rotary Engine", 214
Kasino, Heinkel, 122
Kenworth truck, Boeing gas turbine, 120
KHD. *See* Klöckner-Humboldt-Deutz, 26
   Dz 710 diesel engine, DVL/Hirth turbocharger, 26
Kilpatrick, D.A., 148
KKM (Kreiskolbenmaschine stationary outer housing), NSU/Wankel
  project, 152
   difficulty with apex seals of rotor, 153
   NSU, 181
Klöckner-Humboldt-Deutz (KHD), 26, 228
Knight double-sleeve engine, Daimler Motoren-Gesellschaft, 3
Knudsen, William S., 46
Kolb, Paul, 88, 96
Kombinationsverdichter,
   Heinkel jet engine compressor, 50
   Heinkel jet engine He S, 011, 50
Kronogard, Lycoming AGT 1500 tank gas turbine, 225
Kronogard, Sven Olof, Kronogard Turbine Transmission, 225
   differential turbine principle, 225
   Volvo passenger car, 225
Kruckenberg, Franz, 142
Kutta, Wilhelm, 257

Lear, William P., 219
   Lear Jet, 219
   steam turbine, 219
Leist, Fritz, 63
Lenoir engine, 160
Lewis Flight Propulsion Laboratory, NACA, 146
Lewis Research Center, NASA, 189
LFA. *See* Luftfahrtforschungsanstalt Völkenrode
LHR. *See* Low Heat Rejection
Lift/cruise engines, 171
Lightweight lift-cruise jet engines and turbofans, WAD jet engine
  development, 171
Lilienthal Gesellschaft für Luftfahrtforschung, 2, 21, 60, 139
Lilienthal, Otto, 261
Lilienthal Society, Conference on Ceramic Gas Turbine Components, 60-61
Liquid Metal Regenerator (LMR), WAD, 175-176
Liquid rocket engines, 123
Liquid-cooled engine,

Napier Sabre, 6
Rolls-Royce Eagle, 6
Liquid-cooled supercharger, Rolls-Royce Merlin engine series XLVI, XLVII, and 61, 26
Liquid-cooling,
    reciprocating aircraft engine, 211
    Wankel aircraft engines, 211
Liquid-fuel combustion, He S 3 B engine, 38
Ljungstrom rotary regenerator, 84
    gas turbine, 91
Lockheed, 36, 173, 178
    Constellation passenger airplane, 102
    Orion, 82
Lockheed L-1011 Tristar jetliner, Rolls-Royce RB 211 high-bypass turbofan engine, 178
Locomotives, coal burning gas turbine, 100
L'Orange, Prosper, 29
L'Orange Company, 29, 123
L'Orange "R" engine fuel injection system, "pumpless injection", 29-31
Lorenzen Exhaust Turbocharger, 176
Low compression ratio, stratified-charge, Curtiss-Wright Wankel rotary development, 166-168
Low cycle fatigue, Parsons powerplant failure, 101
Low Pollution Power Systems Development Program (LPPSD),
    alternative fuels, 216
    Brayton cycle (gas turbine), 216
    electric cars, 216
    EPA/NATO, 216
    Rankine cycle (steam engine), 216
Low-heat-rejection (LHR) engine, 216, 226
Low-power gas turbine engines, 195-200
Lowthian, Charles, S., 90, 96-98
Luftfahrtforschungsanstalt Völkenrode (LFA), 83
Lundquist, Wilton G., 81, 123, 137
Lycoming,
    AGT 1500 Abrams tank, 229
    AGT 1500 tank gas turbine Chrysler Corp., 192
    differential turbine, 198
    low-power gas turbine, 195-200
    LTS 101 turboprop engine, 200
    LTS 101 turboshaft engine, 200
    PLT32 engine, 195
    Roto-Lobe engine, 187-189
        auxiliary rotors, 187
        engine geometry, 187
        Marshall Tri-Dyne, 212

power rotor, 187
T53 engine series, 189-193, 198
T55 engine, 200-201
T55 engine series, 189-193, 198
tank gas turbine, 194
Lycoming Stratford Division, AVCO Corp, 187-192, 194-195, 198, 200-201, 203, 206, 212, 222, 229
Lycoming/General Electric, recuperated two-shaft turbine tank engine, 227-228
Lysholm, screw-type compressor, 84

M-1 Abrams Main Battle Tank, AGT 1500 tank gas turbine, 226-228
M-60 tank, 227
Mack Trucks, 228
Madelung, Georg, 122
Mader, Otto, 38
Magnetic levitation (MAGLEV), alternative railroad, 144
Mahle, Ernst, 120-121, 179, 207
Mahle Group, 121
Mahle, Hermann, 120
Mahle Morristown, 208
Mahle piston factory, automotive gas turbine, 120-122
MAN Company, 52, 181
Marine Division, Parsons, 100
Marshall engine, 214
    rotary compressor, 212
Marshall, John, 187
    rotary compressor, 187
    Tri-Dyne engine, 187
Marshall Tri-Dyne, Lycoming Roto-Lobe, 212
Marshall Tri-Dyne rotary engine, 212
Marshall/Roto-Lobe engine, 187-189, 201
Maschinenfabrik G.F. Grotz, Heinkel Group, 80
Massachusetts Institute of Technology (MIT), 218
Mauch, Hans, 44
Maybach Motorenbau, 83
Mazda Motor Corporation, 166, 209-211
    rotary-engined passenger cars, 210-211
    Wankel engine, 211
    Yamamoto, Kenichi, 210-211
Mazda sports car, rotary engine development, Toyo Kogyo, 209
Mercedes, Daimler-Benz, 141, 300
Messerschmitt, 33, 49, 115, 118
Messerschmitt, Willy, 35
Messerschmitt Me 109 piston engine fighter, 49
Messerschmitt Me 262 fighter plane, Junkers Jumo 004A engine, 45
Messerschmitt Me 262 twin-jet fighter, Junkers Jumo 004 engine, 47

Messerschmitt Me 262/Jumo, 004, 49
Metallic gas turbines,
    for bus, General Motors, 222
    for industrial applications, General Motors, 222
    for truck, General Motors, 222
Methanol, alternative fuel, 218
Military applications, stratified-charge Wankel engine, 211
Military pickup trucks, Stirling engine, 220
Military vehicles, M-1 Abrams Main Battle Tank, 226-228
Miller Jet Engine Group, 51
Ministry of Fuel and Power, 99
Ministry of Supply (MoS), 27, 87-88, 94, 96, 99, 102, 105
MIT. *See* Massachusetts Institute of Technology
ML He S 50 experimental engine, Heinkel-Hirth, 63
ML Motorrückstößer-Luftstrahltriebwerk engine jet propulsor, aircraft
    propulsion, 61
MoS. *See* Ministry of Supply
MoS, Fighting Vehicles Research and Design Establishment (FVRDE), 88
MoS/Parsons, 103
    tank gas turbine engine, 103
Moss, Sanford A., 41
Moss supercharger, 46
Moss turbocharger, Boeing B-17 bomber, 27, 61
Muffling, Turbo Compound TC 18, 142
    Curtiss-Wright Research Division, 142
    Helmholtz resonator sound dampers, 142
Müller, Max Adolf, 41, 44, 117
    He S 30 engine, 41-44
Müller-von Ohain, thermal compression gas turbine, 117
Multifuel engine, Curtiss-Wright, Wankel rotary development, 179
Multifuel piston engine, 165
Multifuel rotary engines, 165-168
Multifuel stratified-charge engines, John Deere Company, 210

NACA Lewis Flight-Propulsion Laboratory, conference on air-cooling of
    gas turbines, 146
NAHBE (Naval Academy Heat Balanced Engine), 214
Napier, Nomad, 27-28
Napier Company, 6, 28
    H-Type Sabre engine, 6
Napier Nomad 2, turbocompound diesel engine, 27-28
Napier Sabre, liquid-cooled engine, 6
NASA, 190, 210
    Lewis Research Center, 189
National Air and Space Museum, Session marking the 40th Anniversary of
    the first jet flight, 68

National Automotive History Collection, 243
    Detroit Public Library, 212
NATO. *See* North Atlantic Treaty Organization
Natural gas, alternative fuel, 218
Neumann, Gerhard, 174
Nissan, 194
Noise abatement, 1, 142
Nomad, Napier, 27-28
Norbye, Jan P., 241-242
North Atlantic Treaty Organizaion, (NATO), 216, 223
    Committee on the Challenge of Modern Society (CCMS), 216
    fighter/bombers, Sapphire J 65 turbojet engine, 137
NSU, 113, 151-154, 157-161, 181, 205, 209
    KKM, 181
    Prinz, 154, 181
    Ro80 twin-rotor Wankel engine family sedan, 209
    Spider, 181
NSU scooter, Ebert drive, Hydrostatic axial piston pump transmission, 112-113
NSU/Wankel, 152, 157, 158-160
    anti-friction ball and roller bearing, 157
    circumferential cooling flow, 157
    DKM (Drehkolbenmaschine dual rotation machine), 152-153, 158
    gas sealing network, 157, 159
    KKM (Kreiskolbenmaschine stationary outer housing), 152, 158
NSU/Wankel license, General Motors Corporation, 209

OEM engines, Heinkel, 106-108
Oestrich, Hermann, 47, 83
    BMW, 003, 47
Ohio River Bridge collapse, Roebling, John A., 23
Oldsmobile, 141
Orion, Lockheed, 82
Otto engine, 13, 160, 165, 181, 194-195, 197, 215
Otto, Nikolaus August, Deutz Company, 181
Otto-Diesel engine, 29-32, 120, 165
Otto-Diesel type "R" engine, Hirth-Motoren, 29
Outboard Marine Corp (OMC), 165

Parsons,
    gas turbine development for industrial and marine applications, 102
    Marine Division, 100
    tank engine project, 102
    tank gas turbine, 228
Parsons, Charles A., 24, 88, 94, 96, 99-101, 103, 137, 225

Parsons powerplant failure,
    Conference in Toronto, 101
    low cycle fatigue, 101
    thermal expansion, 101
Passenger car, Chrysler gas turbine, 223
Perfect Circle Company, Curtiss-Wright Wankel rotary development, gas sealing problem, 159
Peugeot Company, 31
Pfister, Magda (Bentele), 1
PIP, 137-139
piston engine, 121, 137, 194
    radial centrifugal supercharger, 13
Planck, Max, 33, 84
PLT 27 turboshaft engine, 197
    UTTAS military helicopter, 192
Pohl, Robert W., 36
Pollution Conference, AVCO, 194
Poppet valve, reciprocating engine, 5-6
Porsche Company, 31
Porsche, Ferdinand, 35
Potts, Matthew, 90
Power turbine, 94, 99
Prandtl, Ludwig, 24, 84
Pratt & Whitney, 148, 173
    twin-spool engine, 148
Prince Philip, Duke of Edinburgh, 96
Princess flying boat, 101
Princeton University, 165
Prinz,
    NSU, 181
Product Improvement Program (PIP), Wright Aeronautical Division (WAD), 137
"Project Smoothie", DAC/CW, 142
Propellertrain, Turbo Compound TC 18, 142-145
PTL DB/He S 021 experimental engine,
    Daimler-Benz/Heinkel-Hirth, 63
PTL Propeller-TL turboprop, Aircraft propulsion, 61
"Pumpless injection", L'Orange "R" engine fuel injection system, 29-31

"R" engine, 29-31
    fuel injection system,
        Bosch, 29-30
        L'Orange, 29-31
R-Engine, Svenska Rotor Maskiner, 212
Radial centrifugal supercharger, piston engine, 13

Radial compressor, 68
  De Havilland Company, 90
Radial flow compressor, 200
Radial-inflow turbine, 49, 198
  A/S Vapenfabrikk Kongsberg KG 2/3, 49
  He S 8 jet engine, 49
Ramjet, 123
  aircraft propulsion, 33
Rankine cycle, steam or vapor engine, 218-220
Rankine cycle (steam engine), Low Pollution Power Systems Development
  Program (LPPSD), 216
RB 211 high-bypass turbofan engine, Rolls-Royce Composite hollow propeller
  blades, 178
RC1-60 experimental engine, Curtiss-Wright Wankel rotary development, 161
RC4-60 experimental engine, Curtiss-Wright Wankel rotary development, 161
Reciprocating aircraft engine, 138
  liquid-cooling, 211
Reciprocating engine, 5-6, 8, 33, 63-64, 84, 123, 137, 151, 157, 164, 187,
  194-195, 197, 221
  aeroacoustics, 142
  poppet-valve, 6
  sleeve-valve, 6
Reciprocating engine flat-disk valve, Wankel, 8
Reciprocating engine/propeller drive, 33
Regener, Erich, 257
Regenerated gas turbine, 84, 194, 197, 216, 222
Regenerated turbine, 222
Regenerator, 94
Reitsch, Hanna, 1
Research Division, Curtiss-Wright, 142
Reverse-flow combustor, Whittle, 68
Ricardo Consulting Engineers, 168
Ritz, Ludolf, 84, 88, 91-92
Ro80 twin-rotor Wankel engine family sedan, NSU, 209
Robinson, S.T., 81
Rocket engine, 215
Rocket propelled He, Heinkel, 117, 176
Rocket-assisted glider, von Opel, Fritz, 33
Rocket-powered He 176, Günter, Siegfried, 82
Rocket-propelled automobile, von Opel, Fritz, 33
Rockets, aircraft propulsion, 33
Roebling, John A.,
  Brooklyn Bridge, 23
  Ohio River Bridge collapse, 23
Rolls-Royce, 6, 26, 68, 178, 192, 195

Composite hollow propeller blades, 178
Eagle engine, 6
Merlin engine series XLVI, XLVII, and 61, liquid-cooled supercharger, 26
RB 211 high-bypass turbofan engine,
    Lockheed L-1011 Tristar jetliner, 178
    Roll-Royce bankruptcy, 178
RS 360 engine, 192
Rotary compressor,
    Marshall engine, 212
    Marshall, John, 187
    Svenska Rotor Maskiner R-Engine, 212
Rotary engine, 149, 151-168, 187, 189, 231
    engine development, 168
        engine geometry, 158, 179, 187
        Toyo Kogyo, 209
    passenger, sports, and race cars, mazda, 210
    Roto-Lobe, 187-189
    session, SAE International Congress, 164
    Svenska Rotor Maskiner AB (SRM), 212
    technology, 182, 205
        Curtiss-Wright, 182
        Yamamoto, Kenichi Mazda Motor Corp, 210-211
    test results, 163
Rotary regenerator, 84, 89, 91-92, 194
    tank gas turbine, 89
Rotary valve engine, 10
    Aspin, 10
    Baer, 10
    BMW, 10
    "Cross cool combustion chamber", 10
Rotating Combustion Engine Development, WAD, 163-164
Roto-Lobe engine, Lycoming, 187-189
Rover, gas turbine-powered automobile, 120
Rover Company, 93, 120, 225
    gas-turbine-driven passenger car, 93
    Vehicular gas turbines, 93, 225
Rover Jet 1, gas turbine-driven passenger car, 221
Royal Aeronautical Society, Conference on radial and axial compressors, 68
Royal Aircraft Establishment (RAE), 81
Royal Automobile Club, 93, 119
Russel, William A., 90
Russian MIG fighters, Gunter, Siegfried, 122

SAAB, 108
SAE International Congress, Rotary Engine session, 164

SAE (Society of Automotive Engineers), 139, 161, 163-165, 168, 209, 211, 219, 226, 241-243
Sänger, Eugen, 122
Sänger, Irene, 122
Sapphire J 65 turbojet engine,
  Curtiss-Wright, 137
  NATO fighter/bombers, 137
  U.S. fighter/bombers, 137
  U.S. Navy Blue Angels F11F-1 "Tiger", 137
SBAC, 118
Schelp, Helmut, 66, 82
Schif, Curt, 14, 16, 21, 48
  Heinkel-Hirth, 48
Schmidt, Fritz F.A., 24-25
Schneider speed-race Trophies, 33
Schnürle, Adolf, 26
Schnürle loop-scavenging scheme, Heinkel OEM engine, 106
Schweinfurter Präzisions-Kugellager-Werke, Fichtel & Sachs (F&S), 257
SCORE (Stratified Charge Omnivorous Rotary Engines), John Deere
  Company, 210
Screw-type compressor, Lysholm, 84
Second Annual Rotating Combustion Conference, Curtiss-Wright Wankel
  rotary development, 179-181
Shackleton, 28
Shannon, J.F., 148
Shelp, Helmut, 44
Siemens & Halske engines, Siemens Group, 1
Siemens Group, Bramo, 1
Sigma Xi, 243
Simpson, Ernest C., 81
Sinclair C5, electric car, 231
Sinclair synchrocoupling, tank gas turbine, 89
Single rotor lift fan engine, verticle takeoff and landing aircraft (VTOL),
176-177
Single shaft jet engine, BMW 003, 47
Single-sleeve engine, Burt and McCollum, 3
Single-sleeve valve engine, 3-4
Single-spool, 200
Single-spool compressor, 198
Single-spool jet engine, 137
  General Electric, 148
Single-stage axial turbine, 192, 198
  tank gas turbine, 89, 192
Skorski, R., 81
Sleeve-valve engine, 4-6
Smith, Geoffrey, 65

Smithsonian Institution, 68
SNCF railroad, 144
SNECMA. *See* Société National d'Etude et de Construction de Moteurs
   d'Aviation, 83
Société National d'Etude et de Construction de Moteurs d'Aviation
   (SNECMA), 83
   ATAR engine, 122
Society of Automotive Engineers. *See* SAE
Society of British Aircraft Constructors, 117
Society of Motor Manufacturers and Traders, 102
Society of the Sigma Xi, Steven Chapter, 165
Sollinger, Ferdinand (Freddie) P., 157
Sonderentwicklung Department (Special Development), Heinkel, 36
Southwest Research Institute, 31
Soviet Ministry of Defense, 178-179
Spider, NSU KKM, 181
"Split Cycle Rotary Engine", Karol engine, 214
Sprengniete (explosive rivet),
   Butter Brothers, 116
   Heinkel-Butter, 116
Sputnik, 171
SRM. *See* Svenska Rotor Maskiner AB
SST. *See* Supersonic transport engine
Stationary recuperators, 91, 194
Steam boilers, coal burning gas turbine, 100
Steam engine, 160, 218-219
   vehicular application, 219
Steam turbine,
   gas turbine technology, 88
   Lear, William P., 219
Stevens Chapter, Society of the Sigma Xi, 165
Stevens Institute of Technology, 165, 218
Stirling engine, 219-220, 223
   bus, 220
   compact car, 220
   Dutch Philips Company, 219-220
   Ford Motor Company, 220, 225
   General Motors, 220
   generating set, 220
   military pickup trucks, 220
   riding mower, 220
   yacht, 220
Stirling engine-driven bus, Dutch Philips Company, 220
Stratified charge with coordinated fuel injection and ignition, Wankel diesel
   application, 166

Stratified-charge, Hesselman engine, 166
Stratified-charge rotary engine, Jones, Charles, 210
Stratified-charge Wankel engine,
  commercial applications, 211
  military applications, 211
Stratocruiser passenger airplane, Boeing, 102
Stuttgart Technische Hochschule (TH Stuttgart), 1, 63
Subway train, flywheel, 232
Supercharged Otto-type engine, turbine cooling, 16
Supercharger, 26-28, 45, 64, 138-139
  axial flow, 13
  gear driven, 13
  radial centrifugal, 13
Supersonic transport (SST) engine, 172-174
  TJ70, 172-174
  WAD, 172-174
Svenska Rotor Maskiner AB (SRM), 212-214
  R-Engine, rotary compressor, 212
Symposium for the 50th Anniversary of the first jet-powered flight, AIAA, 68
Symposium, Alternative Automotive Power Systems (AAPS), 212
Symposium "50 Jahre Turbostrahlflug - 50 Years of Jet-Powered Flight",
  DGLR, 69
Szydlowski, Joseph, 83

T53 engine, Lycoming, 189-193
T53-L-13 helicopter engine, compressor blade failure, 189-191, 201
Tacoma Narrows bridge collapse, von Kàrmàn, Theodore, 22-23
Tank engine AGT, U.S. Army, 221, 1500
Tank engine project, Parsons, 102
Tank gas turbine engine, MoS/Parsons, 103
Tank gas turbine, 89-99, 105, 191, 220, 225-228
  AGT, 191, 1500
  centrifugal compressor, 89
  control system, 99
  FVRDE, 210
  gas turbine technology, 88
  rotary regenerators, 89
  Sinclair synchrocoupling, 89
  single-stage axial turbine, 89
  thermal shock, 96-98
  two-stage axial turbine, 89
  U.S. Army, 103
  U.S. Army Tank Automotive Command (TACOM), 176
  Whittle-type combustion chambers, 89
Tank Museum, Bovington Dorset, 103

Tanks, AGT 1500 tank gas turbine, 226-228
TC 18 Turbo Compound engine, 137
Technische Prüfstand, BASF, 29
Teledyne Continental diesel engine, General Motors Corp., 192
Tempo Matador, 107-108, 1400
    Heinkel OEM engine, 106-108
Texaco Combustion Process (TCP) stratified charge, Barber, Everett M., 166
Texaco engines, low compression ratio, 168
TGV. *See* Train à grand vitesse
TH Stuttgart (Stuttgart Technische Hochschule), 1, 63, 141, 257-258
Thermal compression gas turbine, Müller-von Ohain, 117
Thermal expansion, Parsons powerplant failure, 101
Thermal fatigue, vibration fatigue, 98
Thermal shock, 96, 98
   tank gas turbine, 96-98
   tests, 98
      gas turbine materials, 98
   thermal stresses, 96
TJ60 engine, 172
   rig compressor, WAD, 174
   rig engine, compressor blade failures, 175
TJ70, Supersonic transport engine, 172-174
Topfschaufel,
   He S 011, 57
   Turbine blade, 24-25, 57
Tores-Schluss-Panik, 158, 161, 209
Toroidal traction drive,
   Austin Co. Hayes transmission, 177
   automotive and industrial applications, 177
   Beier transmission, 177
   WAD, 177
Touring scooter, Heinkel, 108-116
Toyo Kogyo Co., 166, 209
Train à grand vitesse (TGV), 144
Transmissions, Beier transmission, 112
Transpiration cooling,
   industrial turbines, 175
   stator and rotor blades, 172
   turbine blades, 172
      Curtiss-Wright, 147-148
      WAD, 175
   turbine cooling, 147-148
   variable area turbine, WAD J65 rig engine, 174
   WAD, 175
Transportation History Foundation, University of Wyoming, 174, 241-242

Tri-Dyne engine, Marshall, John, 187
Truck diesel engine, vapor engine, 219
Turbine, 198
Turbine blade,
    Faltschaufel, 24-25, 57
    Topfschaufel, 24-25, 57
Turbine blade failure,
    frequency/resonance/vibration fatigue, 17-21
    Vertol Chinook helicopter Lycoming T55 turboshaft engine, 200
    vibration fatigue, Junkers Jumo 004 B engine, 45-46
Turbine cooling, 16-17, 23-26, 146
    air-cooled blade, 146
    convection cooling, 146-149
    DVL Institut für Motorische Arbeitsverfahren und Thermodynamik, 24
    Faltschaufel (folded blade), 24-25
    folded blade, WMF Topfschaufel (tubular or bootstrap blade), 24
    supercharged Otto-type engine, 16
    technology, 137
        hollow air-cooled blades, 146
        J65 engine, 146
    transpiration cooling, 147-148
Turbine rotor blade failures, Junkers Jumo 004B, 45
Turbine rotor cooling, 121
Turbines,
    advanced gas turbine engines, 148
    AGT 1500 tank gas turbine, 191-193
    air-cooling, 146
    ceramic turbine blades, 222
    development of tank gas turbine, 88-99
    differential turbine application, 198
Turbo Compound TC 18,
    engine, Curtiss-Wright, 6
    Moss supercharger, 138
    muffling, 142
    propellertrain, 142-145
    supercharger, Curtiss-Wright, 27
Turbo lag, 14-15
Turbocharged automobile engines, 15
Turbocharged diesel engine, 219
Turbocharger, 25-28, 51, 64, 66
    air-cooling, 146
    development, 82
        diesel engine, 83
    DVL, 14
    exhaust gas turbine, 13
    for reciprocating aircraft engines, Hirth Motoren Company, 33-34

Hirth-Motoren, 14-15
technology, 208
Turbocompound diesel engine, 27-28, 219, 229
Napier Nomad, 2, 27
Turbojet, aircraft propulsion, 33, 38
Turbojet back to back radial compressor-turbine rotor, von Ohain, Hans, 35
Turbojet TL Turbinen-Luftstrahltriebwerk, aircraft propulsion, 61
Turbomachines, 103
Turbomeca, 83
Turboprop engines, 175
Turboshaft engines, 189-193
Twin-rotor turbocharged stratified-charge aircraft engine, John Deere
 Company, 210
Twin-spool, 200
engine, Pratt & Whitney, 148
jet engines, 123
jet and ramjet engine, Curtiss-Wright, 148
turbomachinery, AGT 1500 Lycoming tank gas turbine, 191-193
Two-stage axial turbine, tank gas turbine, 89

U.S. Air Corps, 81
U.S. Air Force, 138, 171-172, 175, 178
WAD, 172
U.S. Army, 161
tank engine AGT, 221, 1500
tank gas turbine, 103
U.S. Army Tank Automotive Command (TACOM), 176
tank gas turbine, 176
U.S. Congress, 243
U.S. fighter/bombers, Sapphire J 65 turbojet engine, 137
U.S. Government, 220
U.S. Marine Corps, 223
U.S. Navy, 38, 81, 138-139, 147
Blue Angels F11F-1 "Tiger", Sapphire J 65 turbojet engine, 137
U.S. Patent, Wankel engine, Curtiss-Wright, 153-154
U.S. turbine cooling technology, 146
University of Wyoming, Transportation History Foundation, 174, 241
UTTAS, 192
military helicopter, PLT-27 turboshaft engine, 192

Vapor engine, truck diesel engine, 219
Variable-area turbine, 174
VDI. *See* Verein Deutscher Ingenieure
Vehicle engine, 96
Vehicle gas turbine, gas turbine technology, 88
Vehicular application, steam engine, 219

Vehicular gas turbine, 91, 93, 120, 191, 227-228
    Boeing truck, 225
    Chrysler Turbine car, 225
    engine, 208
    Rover passenger car, 93, 225
    technology, 176
Vehicular and industrial gas turbines, 101
Verein Deutscher Ingenieure (VDI), 139, 160, 258
Vertical takeoff and landing aircraft (VTOL), 159
    Curtiss-Wright, 159
    GE/Ryan, 176
    single rotor lift fan engine, 176-177
    WAD, 176-178
Vertol Chinook helicopter Lycoming T55 turboshaft engine, turbine blade failure, 200
Vertol/Bell X-22 Osprey, Curtiss-Wright X-19 experimental aircraft, 178
Vibration fatigue,
    compressor blades, 171-172
        of the lift-cruise engine, 171
    thermal fatigue, 98
    turbine blade failure, 17-21, 45-48, 81
        BMW 003 engine, 47-48
        DVL/Hirth turbochargers, 16-21
        gas turbine compressor, 148
        He S 011 jet engine, 54, 57-61
        He S 8 A engine, 49
        Junkers Jumo 004B, 45-46
Vidal & Sohn, Tempo Werk, 106
Volvo Company, 159, 225
Volvo passenger car, Kronogard Turbine Transmission, 225
von Bach, Carl, 257
von Buz, Heinrich, 181
von der Nüll, Werner, 14, 16, 19, 21, 82
von Eyth, Max, 262
von Heydekampf, Gerd Stieler, 152-154, 158, 181, 205, 209
von Kàrmàn, Theodore, Tacoma Narrows bridge collapse, 22-23
von Laue, Max, 84
von Ohain, Hans,
    He S 8 A engine, 48
    He S 3 B engine, first jet propelled flight, 35-41
    Heinkel Sonderentwicklung (Special Development) Department, 36-37
    turbojet back-to-back radial compressor-turbine rotor, 35
von Ohain, Hans-Joachim Pabst, 35-38, 49, 66-68, 117, 181
von Opel, Fritz, 33
    rocket-assisted glider, 33
    rocket-propelled automobile, 33

Voysey, Reginald, 81, 91
VTOL. *See* Vertical takeoff and landing aircraft
Vulcan delta-wing bomber, AVRO, 118

WAD (Wright Aeronautical Division), 123, 138, 147-148, 153, 158, 163, 165, 174-177, 179, 191
    aviation and aerospace design studies, 177
        astronaut "cap pistol", 177
        "diluent stator", 177
        VTOL, 177-178
    flexible compressor blades/friction dampers, 175
    J65 rig engine, transpiration-cooled variable area turbine, 174
    jet engine development, 171
        lightweight lift-cruise jet engines and turbofans, 171
        lightweight lift-cruise turbojet TJ60, 171
    lightweight gas turbine engines, 171-175
    liquid metal regenerated turboprop, 175-176
    Liquid Metal Regenerator (LMR), 175-176
    production aircraft engines and jet engines, 158
    propellertrain, 158
    rotating combustion engine development, 163-164
    supersonic transport engine, 172-174
        TJ70, 172-174
    TJ60 rig compressor, 174
    toroidal traction drive, 177
    transpiration-cooled turbine blades, 175
    transpiration cooling, 175
    U.S. Air Force, 172
    verticle takeoff and landing aircraft (VTOL), 176-177
Wagner, Herbert, 41
Wagner-Müller-Heinkel, He S 30 engine, 41-44
Wankel, 153-154, 158-161, 205, 209-210, 241
    DKM-54 engine, 153-154
    reciprocating engine flat-disk valve, 8
Wankel aircraft engines,
    air-cooling, 211
    liquid-cooling, 211
Wankel diesel application,
    airflow pattern, 166
    stratified charge with coordinated fuel injection and ignition, 166
Wankel engine, 151-168, 181, 187, 195, 197, 205, 209, 212, 242
    compression ratio, 165-166
    Curtiss-Wright vertical takeoff and landing airplanes (VTOL), 159
    development, Curtiss-Wright, 151
    General Motors, 220
    Mazda, 211

Wankelwalze sealing element, 8
Wankel, Felix, 8, 151-152, 161, 166, 168, 181, 212-214
    von Heydekampf, G. Steiler, NSU, 181
Wankel II engine, 212-213
Wankel Institute, 8, 168
Wankel rotary engine, 144, 211
    apex gas seal problem, 209
    development, 151-152, 164-165
        combustion problem, 151, 153
        cooling problem, 151, 157
        heavy fuels, 210
        ignition problem, 151
        main bearing problem, 157
        oil and gas sealing problem, 151, 156-157
    diesel applications, 165
        difficulties,
            combustion, 166
            combustion chamber shape and gas seals, 166
            gas sealing, 166
            mechanical, 166
        starting and low speeds, 166
Wankelwalze sealing element, Wankel engine, 8
Ward's Engine Update (WEU), 226
Ward's Wankel Report (WWR), 217
WAS, 172
Water-cooling of gas turbines, 100
    external cooling, 100
    internal cooling, 100
Weinrich, Helmut, 47
Weise, Arthur, 122
West German Federal Ministry for Research and Technology (BMFT), 223
Westinghouse, 222, 225
    ceramic turbine, 222
Weyl, A.R., 118
Whittle, Frank, 28, 65-68, 118, 181
    jet engine, 28
    reverse-flow combustor, 68
Whittle-type combustion chambers, tank gas turbine, 89
Wilke, Wilhelm, 29
    BASF Otto-Diesel aircraft engine, 29
Williams, Calvin, 218
Williams, Charles, 218
Williams Corporation, 223
Witzky, Julius E., 31
WMF. *See* Württembergische Metallwarenfabrik

WMF Topfschaufel (tubular or bootstrap blade),
    Braig, Hans, 24
    Turbine cooling folded blade, 24
WMF Topfschaufel air-cooled turbine blades, 57-61
Wolff, Harald, Heinkel-Hirth jet engine development, 48
Wright Aeronautical Division (WAD),
    Curtiss-Wright Corporation (CW), 81, 123, 137
    Product Improvement Program (PIP), 137
Wright,
    Cyclone aircraft engine, 123
    piston engines, 137
    turbojet, 137
Württembergische Metallwarenfabrik (WMF), 24
WWR, 211-212, 214

XAMAG, 206

Yamamoto, Kenichi, 209-211
    Mazda Motor Corp, 210-211
        rotary engine technology, 210-211
Yanmar Diesel Co., 166

Zadnik, Otto, 88, 96
ZTL DB 007 experimental engine, Daimler-Benz, 63
ZTL Zweikreis-TL turbofan, aircraft propulsion, 61